James Cannon specialises in organisation design and development and has consulted with organisations in the UK, North America, Europe, Middle East, Africa and Asia. He works as a coach to directors and senior executives, as well as mediating and facilitating events designed to improve the effectiveness of Boards and organisations. He speaks regularly to directors and at conferences and ran an extensive range of training courses for the Chartered Institute of Personnel and Development (CIPD) in the UK, as well as for other organisations. He was a visiting lecturer at Geneva University.

He was a fellow of two institutes (CIPD and CMI), British Psychological Society member and Manpower Society prize winner. He was formerly special adviser to the CIPD. In 2016, he gained an award from the British Government for his work in leadership development.

With a degree in Behavioural Science, a master's degree in Manpower Studies and a doctorate in Organisational Psychology, his career has included director appointments at Charterhouse, Black and Decker and Thorn EMI, in Organisation Development, Human Resources, Information Technology and General Management, including founding and becoming Chief Executive of Solutions Electronic Services.

In 1989, he started his own consultancy, as well as co-founding Cavendish Partners, now merged with Right Management, a firm specialising in career counselling and coaching for senior executives. He is a trustee of Trinity, a charity working with homelessness in London; and of the Tracy Trust, a charity helping pensioners in Buckinghamshire.

He has written several books, including *Cost-Effective Personnel Decisions, Database Directory, Giving Feedback, Making the Business Case, Talent Management and Succession Planning, Organisation Development and Change* (the latter two with Rita McGee), *Organisation Design and Capability Building* (with Rita McGee and Naomi Stamford) and workbooks such as *Team-Based Problem-Solving* and the *Career Review workbook*. He was consulting editor of *Toolclicks*, the online service from CIPD.

Robert Cannon

Robert Cannon is a teacher, coach, facilitator and educator specialising in the arts and entertainment industry. He draws upon two decades of experience on the frontline of the music business in his work with the organisations and practitioners at the forefront of the arts and entertainment industry, as well as with the students who are its future. He has worked in the UK, USA, Australia and throughout Europe.

Rob is currently an academic lecturer in Arts & Entertainment Management at the Australian Institute of Music (AIM), having previously served as head of school and overseeing the degree programmes in Arts and Entertainment Management, Composition and Music Production, Audio Engineering and Dramatic Arts. Rob continues to write and teach a broad variety of courses and classes, including Creativity, Marketing, Strategic Partnerships, Introduction to Arts and Entertainment Management, Performance Psychology and others.

Rob has spent 20 years working in music. His first brush with the music industry was as a guitarist and pianist in several rock and jazz bands in the UK, before he started his career on the business side of the industry at Clive Davis' J Records in New York.

Rob subsequently worked in A&R and marketing for several Sony and BMG labels in New York, London and Sydney, working with artists ranging from Rod Stewart to Alicia Keys to Maroon 5.

Rob has coached individuals within the arts and entertainment industry and has designed workshops for various music and entertainment companies. He has authored and presented talks and conference papers on teaching and harnessing creativity within the arts, and on applying techniques from the field of positive and performance psychology within the arts. He has also appeared on panels at various music industry conferences, and has written about music, the arts and travel for a range of publications, websites and newspapers.

Rob holds a Bachelor of Arts in Modern Languages from Cambridge University, a Master of Arts in Music Business from New York University and a Master of Applied Science in Coaching Psychology from Sydney University.

To every person who has had an 'aha! moment' but is yet to share it with the world.

Robert Cannon and James Cannon

AHA! A USER'S GUIDE TO CREATIVITY

Inspiring Insight and Innovation in
Individuals and Organisations

AUSTIN MACAULEY PUBLISHERS™

LONDON * CAMBRIDGE * NEW YORK * SHARJAH

Copyright © Robert Cannon and James Cannon 2023

The right of Robert Cannon and James Cannon to be identified as authors of this work has been asserted by them in accordance with sections 77 and 78 of the Copyright, Designs and Patents Act 1988.

All rights reserved. No part of this publication may be reproduced, stored in a retrieval system, or transmitted in any form or by any means, electronic, mechanical, photocopying, recording, or otherwise, without the prior permission of the publishers.

Any person who commits any unauthorised act in relation to this publication may be liable to criminal prosecution and civil claims for damages.

The story, experiences, and words are the author's alone.

A CIP catalogue record for this title is available from the British Library.

ISBN 9781398425484 (Paperback)
ISBN 9781398412613 (ePub e-book)

www.austinmacauley.com

First Published 2023
Austin Macauley Publishers Ltd®
1 Canada Square
Canary Wharf
London
E14 5AA

For Rob, thanks are due to the Australian Institute of Music for asking him to develop a course on *creativity* a decade ago and allowing him to teach it ever since. Huge thanks to the many students who've taken an interest in these classes and challenged him to think deeply with their questions, ideas and perspectives over the years. Thanks to his colleagues at Sony Music, and the artists and their teams that he has worked with, for teaching him valuable lessons in creativity and giving him the opportunity to explore various theories and techniques for enhancing it. And finally, thanks to Beth Krzyzkowski Cannon for her unwavering support, and to Sue, Clare and Ed for their encouragement.

For James, many people have contributed, often unwittingly, to the development of the ideas in this book.

His thanks are due to Rita McGee, Roger Niven, Neil Johnston, Richard Coates, Tim Harding and Sue Cannon for helpful insights. He is indebted particularly to the many students at courses and conferences around the world for their challenging questions, which have sharpened his own thinking.

He has had the privilege of working with many clients and organisations over the years and he is grateful to them for the opportunity of learning, as well as contributing. Many of the examples in the book come from these experiences. However, what deficiencies there may be are all his.

Table of Contents

Introduction An Overview — 15
    Why Is Creativity Such a Hot Topic? — 17
    Why Is Creativity So Challenging? — 19
    Our Approach — 23
    About this Book — 24

Chapter 1: Creativity, Innovation and Problem-Solving — 26
    What Is Creativity? — 26
    A Definition of Creativity — 27
    Novelty — 29
    Value — 31
    Creativity — 33
    Innovation — 33
    Problem-Solving — 37
    The 'Aha!' Moment – A Process Model of Creativity — 37
    Chapter Summary and Conclusions — 41

Chapter 2: The Causes and Components of Individual Creativity — 42
    Introduction — 42
    The Underlying Principles of Creativity — 51
    Chapter Summary and Conclusions — 61

Chapter 3: Enhancing Our Individual Creativity — 62
    Introduction — 62

The Personal Conditions for Creativity	*62*
The Search for New Ideas	*70*
Observing the World Around Us	*71*
Learning	*80*
Chapter Summary and Conclusions	*86*

Chapter 4: Navigating and Overcoming the Blocks and Barriers to Creativity — **87**

Introduction	*87*
The Blocks and Barriers to Achieving Our Creative Goals	*89*
Grit	*91*
Mental Toughness and Resilience	*93*
Use CBT Approaches to Address the Fear of Failure and Negative Self-Talk	*97*
Reframe a Fear of Failure	*98*
Value and Praise the Effort, Not the Outcome	*99*
Goal-Setting: Smart Goals, Process Goals and Sundowners	*100*
Other Techniques	*102*
The Role of the Supportive or Critical Friend	*102*
A Caveat	*103*
Chapter Summary and Conclusions	*103*

Chapter 5: Techniques for Idea Generation — **105**

Introduction	*105*
Chapter Summary	*109*

Chapter 6: Creating an Organisational Culture for Creativity and Innovation — **170**

Introduction	*170*
The Challenge of Organisation Design	*171*

Creative Cultures	*173*
Leadership	*178*
Organisation Design – Job Design, Structure and Process	*198*
Creating Innovative Teams	*205*
Rewarding Innovation	*217*
Resource Allocation and Control	*220*
Architecture and Building Design	*221*
Helpers and Hinderers for an Innovative Culture	*224*
Chapter Summary and Conclusions	*227*

Chapter 7: Designing Problem-Solving Workshops and Other Creative Events — **228**

Introduction	*228*
Setting Up a Group Process	*229*
A Generic Problem-Solving Process	*235*
A Note on Idea Management Systems	*240*
Other Approaches to Problem-Solving	*241*
The Osborn Parnes Creative Problem-Solving Model (5)	*241*
The Basadur Simplex Model (6)	*244*
Team-Based Problem-Solving	*245*
Chapter Summary and Conclusions	*256*

Chapter 8: From Creative Spark to Acceptance — **257**

Introduction	*257*
1. Process	*258*
2. Persuasiveness	*263*
3. Confidence	*270*
4. Commitment	*272*
5. Practicality	*275*

6. Working with Resistance	*282*
Chapter Summary and Conclusions	*286*
Chapter 9: Creative Societies and Cities, and a Look to the Future	**287**
Introduction	*287*
External Factors of Creativity	*287*
How Ideas Spread	*290*
The Creative Class and Creative Societies	*294*
Considerations for Education	*296*
Legal and IP Protection	*301*
National Initiatives	*303*
Technology, Artificial Intelligence and Neuroscience	*305*
The Future	*307*
Appendix	**309**
Uses for a Paper Clip	*309*
Broken Squares Game	*309*
A Gantt Chart	*310*
References and Bibliography	**311**

Introduction
An Overview

"Eureka!" Archimedes cried as he jumped out of the bath. The Greek philosopher was so taken by his sudden insight into hydraulics that he proceeded to run down the street stark naked, loudly proclaiming his discovery and totally oblivious to his lack of clothes. After working on various mathematical theories for a lengthy period of time, he had finally experienced the flash of insight—the 'aha!' moment—that brought the solution he'd been searching for. His discovery changed the nature of engineering, and its monumental impact on civilisation has been felt ever since.

One morning, Paul McCartney woke up with a melody in his head. As he sang it to himself, he used the words 'scrambled eggs' to accompany the tune until he could write some more meaningful lyrics. When he finally did so, that simple melody that popped into his head became the Beatles song *Yesterday*, one of the most successful and most-covered songs of all time.

Grace Hopper was on the team at Harvard that invented the Mark 1, one of the first computers. One day she realised that programming computers would be much easier if programmers could use English, and then rely on the computer itself to translate this English into machine code. She was told repeatedly that it couldn't be done, but she persisted and eventually succeeded in inventing the compiler—the program that did the translation—herself.

After further hurdles, borne out of people's refusal to believe that her idea would actually work, the approach was finally widely adopted and became the precursor to the computer languages that heralded the explosion in computing. Incidentally, while investigating an issue in the Mark II computer at Harvard, Hopper and her team came across a moth that was stuck in the machine and jamming the electrical switches—a literal computer bug. So they had to 'debug' the machine to get it working again. The moth has been preserved to this day, taped into a logbook that is now held at the Smithsonian Museum.

Throughout the history of human existence, a small number of individuals like Archimedes, Paul McCartney and Grace Hopper have expressed new ideas or made discoveries that have significantly impacted the world. They are canonised in the fields that they changed forever—science, business, the arts—and their influence across society has been palpable and long-lasting.

We are familiar with the stories of many of these so-called geniuses and their creative brilliance, frequently focussing on the details of the 'aha!' moment: Archimedes in his bath; Isaac Newton articulating his theory of gravity after observing an apple falling from a tree; Lin-Manuel Miranda dreaming up the musical *Hamilton* whilst relaxing on a Mexican beach holiday.

Yet these stories reveal only a small part of the picture, and how these flashes of insight and creative brilliance actually occurred remains something of a mystery.

Meanwhile, on a daily basis, individuals the world over—mere mortals in the realm of creative genius—are also experiencing their own creative breakthroughs to solve the everyday problems that they face. Whether it's dreaming up 'life-hacks' to fix a broken fan with an elastic band or to find a lost earring by using a pair of tights stretched over a vacuum cleaner (1), or coming up with novel solutions to everyday problems, such as taking a new route to a destination when the highways are closed, people exercise their capacity to be creative as they come up with these novel solutions.

Scientist and inventor Genrich Altshuller (2) was imprisoned for political reasons in the former Soviet Union and was tortured by being kept awake. His response to this unfamiliar and unsavoury situation was to put pieces of paper over his eyelids with eyeballs drawn on them. Whenever the prison warden looked into his dimly lit cell, he saw eyeballs and assumed Altshuller was awake. Meanwhile, Altshuller was able to sleep behind the paper, and better withstand the torture. His idea didn't contribute a new insight, discovery or invention to the world, but it did provide a novel solution to the individual problem he faced, with significant consequences. (Altshuller subsequently founded the creative technique TRIZ, which we will discuss in Chapter 5.)

How, then, does creativity work? All these scenarios display creativity in action—albeit in different ways—and also demonstrate the enormous breadth of ground covered by the concept of creativity.

Since the mid-20th century, researchers and academics have begun to approach this question increasingly seriously, exploring it from a number of perspectives. Some focussed on case studies of iconic geniuses and examined various aspects of these individuals, from the weight of their brains to the specifics of their personality traits, and deliberated whether there was anything special about them that could be identified as the source of their genius.

Others focussed on the creative output itself—what were the characteristics of the idea or invention that made it creative? Others looked at the creative process—what were the necessary steps in dreaming up and delivering creative output? And finally, there were those that considered the broader context of creativity—the socio-cultural and contextual elements, the surroundings, and environmental factors for the creative endeavour.

These singular perspectives all turn up various answers and provide overlapping theories on certain aspects of creativity. But the notion of creativity is such a broad and complex entity that understanding it requires more than these perspectives in isolation can offer; a cross-disciplinary approach is needed. We will discuss this further in Chapter 1. But why should we care to know?

Why Is Creativity Such a Hot Topic?

Throughout the ages, creativity has been conceptualised in different ways, and the value placed upon it has varied accordingly. The ancient Greeks and Romans believed that creativity, as exercised in the arts, was delivered by one's muse (known as a 'daemon' to the Romans, a 'genius' to the Greeks).

Creativity has always been instrumental in the arts, but the human capacity to generate creative solutions to everyday problems—previously unrecognised as 'creativity' as such—has also been crucial to our surviving and flourishing. However, we now live in an age where creativity is recognised and highly valued in many other areas of life, not least in business.

In 2010, computing giant IBM conducted a global study across over 60 countries and 200 organisations in order to explore the most desirable traits for leadership in the future (3). The research found that 'creativity' was the most desirable quality for leaders in order to successfully navigate the next five years. A further IBM study in 2012 (4) then found that successful employees were expected to be "collaborative, communicative, flexible, and creative".

A meeting of the World Economic Forum in 2016 discussed the skills that were required to thrive in the 'fourth industrial revolution', and concluded that "creativity will become one of the top three skills workers will need" (5).

Meanwhile, research by urban theorist Richard Florida over the first two decades of the 21st century (6) has estimated that some 30% of the US workforce are part of the 'creative class'—that is, their employment requires creative thinking—and this percentage continues to grow. Evidently, it is becoming increasingly clear that creativity is one of the top requirements for successful leaders, entrepreneurs, business executives…and just about anyone else in a business setting.

The rate at which businesses are overcome by disruptive technologies appears by some standards to be accelerating—from the disappearance of video/DVD stores, to the upheaval in taxi services brought about by ride-sharing apps. In 1960, a company remained on the Standard and Poor's 500 list, which ranks the top 500 companies by size, for an average of 55 years.

By 2017, that timeframe had dropped to 18 years. It took 38 years for radio to reach an audience of 50 million listeners, and only 4 years for the internet to do the same. In Facebook's first year of public access, 200 million people signed up for an account.

The working world is ever-changing, with robots becoming increasingly ubiquitous in manufacturing and elsewhere, and artificial intelligence and Big Data disrupting many industries. The future for individuals and businesses alike is in jobs that require creative thinking, and in ideas for products and services that do not currently exist.

Some 10% of the jobs advertised in the New York Times did not exist only 5 years ago (7). According to manufacturing company Siemens, 50% of their products were not in the product range 5 years ago. The demands of the global marketplace can only be met by agile organisations and agile people.

Such dramatic shifts require creative solutions at every point, not only in products and services, but also in the form of employment and in the way that people and organisations relate to society.

As a result, there has been an explosion of interest in creativity over the past couple of decades, and there are a vast number of books and resources on the topic, from those deeply rooted in science and academia, to those based on anecdotal wisdom and personal experience.

As well as synthesising these multiple perspectives into a cross-disciplinary approach, this book aims to provide a practical approach to the topic, one that is backed up by the science in the field.

Why Is Creativity So Challenging?

There remains a lot of uncertainty around what creativity really is. And this lack of comprehension is not helped by a series of myths and misunderstandings about creativity and the creative process that continue to persist. The beliefs that people are more creative when working alone, or that some of us are born more creative than others, or that creative ideas all emerge from the unconscious in a flash of inspiration, do us no favours in attempting to understand and better harness our creative capacity.

Furthermore, the problem for many of us is that we don't really know what we mean when we use the word 'creativity'. We lavish creative accolades on stars of business, sport, the arts, and praise their ideas and achievements, yet we don't really understand *why* we label someone or something creative. What are the specific criteria we are using? And how do we identify the ingredients and processes behind the creativity in question?

For example, Steve Jobs of Apple is frequently lauded as one of the great creative geniuses of our time. Is that because of the great commercial success of Apple? (If so, why are other successful business leaders not viewed as being equally creative?)

Is it because he came up with a way to change the computing world (even though Apple was not the first to incorporate many of these ground-breaking ideas—Xerox pioneered many elements of desktop computing that Apple was credited for)?

Is it because of the design features on an Apple product (but perhaps that was actually the creative handiwork of Apple designers such as Jony Ive)? When we really try to unpick the process, it becomes very challenging to clearly articulate the pertinent factors.

Creativity is a complex phenomenon to understand because we use this one word to describe so many processes that appear very different, for example:

- solving a brain teaser
- finding multiple uses for a paper clip (a common 'creativity' test)
- finding a word that connects two different objects
- finding a new or unexpected solution to an everyday problem
- envisioning and launching a product that changes our daily behaviours (such as an iPhone or an app like Instagram)
- launching a globally successful company
- discovering a new scientific theory
- penning a novel that gains rave reviews and perhaps even commercial success
- designing an advertising campaign for a new product launch

Various attempts have been made to classify all these different types of creativity. In 1959, for example, creativity researcher Calvin W. Taylor identified five different types of creativity, which he labelled *expressive, productive, inventive, innovative, emergentive* (8). Other researchers have made the distinction between those creative ideas that appear new and creative to the individual who had them (e.g., a solution to an everyday problem that the individual encounters for the first time) but have also most likely been expressed by others previously, and those ideas that are deemed new and creative to a society as a whole, as well as to the individual (e.g., a ground-breaking invention).

Yet despite the various different ways that creativity manifests across all these examples, and the attempts to codify these differences, we still have a tendency to group them together under the one collective term ('creativity'), and then expect them to operate in the same way. No wonder confusion abounds!

We also have a tendency to talk societally about creativity in a way that is frequently myopic and underpinned by a number of limiting or misleading beliefs that are particular to contemporary Western culture.

According to Keith Sawyer (9), an academic who has comprehensively analysed the research on creativity, there are ten prevailing beliefs in the western cultural model of creativity, some of which contain some element of truth, but which in many instances are false. Throughout this book we aim to shed light on the reality of the creative process and provide some tools to navigate that process effectively. The ten beliefs are:

1. The essence of creativity is the moment of insight
2. Creative ideas emerge mysteriously from the unconscious
3. Creativity is more likely when you reject convention
4. Creative contributions are more likely to come from an outsider than an expert
5. People are more creative when they're alone
6. Creative ideas are often ahead of their time
7. Creativity is a personal trait
8. Creativity is based in the right brain
9. Creativity and mental illness are connected
10. Creativity is a healing, life-affirming activity (which is probably correct!)

We canonise individuals who happened to be in the right place at the right time, and often attribute to them ideas they did not actually have. For example, Thomas Edison is frequently credited for inventing the lightbulb, while Henry Ford gets the credit for inventing the motorcar. Both made ground-breaking advances in their respective product, but neither was responsible for their initial invention.

We are prone to believe wayward self-reporting on how the creative process was experienced—tales of creative practice told time and time again by revered artists, inventors and entrepreneurs themselves, in which inspiration and insight frequently strike from the blue and gift the recipient a masterpiece of creative work.

Such stories do not help us understand the process very well; instead, they paint a picture that is at once inspiring and alluring, and yet utterly unattainable to us 'mere mortals.' Charles Darwin claimed in his autobiography to have had a flash of inspiration that brought him his theory of evolution by natural selection.

However, his journals actually show that he played around with various ideas and theories, honing them and refining them over a long period of time, and experiencing a number of much smaller insights, until he arrived at a way to put the ideas together in a theory that made him famous. Author Steven Johnson (10) calls this process the 'slow hunch' and this stands in stark contrast to a singular flash of inspiration in an 'aha!' moment.

We tend to focus on single events in the creative process for specific individuals and theorise that it must be the same for everyone in all creative endeavours. What was the 'genius' doing the morning they had the idea? If we all did that, would we be more creative ourselves? (It's unlikely!)

We over-emphasise certain elements of the process—particularly around the moment of insight—and ignore other important aspects of the creative process. For example, we focus on what Ed Sheeran does in a song-writing session but overlook his years of building guitar-playing expertise that is vital to this process.

What we hear in these tales of creative genius is only ever part of the story. The flash of insight—something most (if not all) of us have experienced at some point or another—can certainly be a part of the creative process. But it is usually preceded by a huge amount of hard work, research, experimentation, collaboration, and so on, that is frequently overlooked; these aspects of the process are certainly omitted or grossly underplayed in the well-worn stories of Archimedes, Newton, Darwin, and many other stories of creative endeavour.

Sometimes, the creative process doesn't even include a flash of insight at all. This is what Arne Dietrich (11) calls the 'trial and error' model of creativity. The prolific inventor Thomas Edison, a man frequently described as a creative genius, certainly showed his commitment to hard work in his invention of a commercially sustainable lightbulb. There was no sudden flash of inspiration or insight in which he miraculously envisioned the material to try for a filament.

Instead, it appears that he experimented with every material he could get his hands on—he tried over 600 materials as light bulb filaments in his experiments in total until carbonised thread turned out to provide him with the results he was looking for[1]. It was hard work that yielded the creative solution; as he himself famously said: "Genius is 1% inspiration, 99% perspiration."

And because generating creative output often requires hard work, discipline, and perseverance, we're all too happy to accept the myths that are paraded and put forward as an excuse to be lazy. As creativity researcher Robert Sternberg (12) argues, to be creative, you must first *decide* to do so—to "generate new ideas, analyse these ideas, and sell the ideas to others".

With these disparate viewpoints, we create a mythology around those who create, and a widespread belief that not all of us can be creative, not all of us will

---

[1] Different versions of this story claim that he experimented with materials ranging from cotton to Chinese bamboo until he found a material that could be carbonised and pass a current. The number of experiments in some accounts rose as high as 5,000 attempts.

receive these creative insights emerging suddenly from our unconscious, not all of us have the potential to bring to life the creative ideas that lead to innovation. And this is a very problematic perspective to hold. No wonder so many of us struggle to produce creative work—to 'be creative'.

Our Approach

As authors of this book, we believe that the idea that only some people can be creative is *not* the case. As we'll explore in this book, we believe that the capacity for creativity is a universal human attribute—all humans have the capacity to be creative in some fashion. We believe that it is a misunderstanding of the creative process, and a resulting inability to bring the ingredients of creativity together, that gets in the way.

A number of commentators point to the expressions of creativity that are so frequently displayed by children, and hypothesise that Western education systems inhibit our creativity by imposing standardised testing, and focussing on 'right' and 'wrong' ways of doing things.

Educator and author Sir Ken Robinson points to a famous study by Beth Jarman and George Land (13) in which they administered a test to measure creative potential—originally developed for Nasa—to a series of children, repeating the testing in a longitudinal study as the children grew older.

What they found is that the percentage of children who tested at the level of genius potential declined as the cohort got older, with the hypothesis being that social acculturation, and education in particular, were supressing and inhibiting the skills, thinking patterns, and use of imagination, that are crucial to creative thinking and creative outcomes.

This theory has subsequently been questioned; whether it is actually the case or not, observing children certainly highlights some of the vital components in the creative process that are accessible to us all, including curiosity, imagination, and play.

About this Book

This book is designed for individuals and organisations that want to understand creativity, the processes that underpin it and the tactics for enhancing it. The book aims to shine a light on what the creative process is really all about…and then to provide a series of tools to help YOU be more creative.

It aims to critically appraise the various theories about creativity, discuss the science around it in an effort to demystify the process, but also then to provide a practical guide to encouraging creativity. We will examine the range of theories about creativity that emerge from studies in psychology, business management, the arts and entertainment industries, along with examples and anecdotes that illustrate the theories in action.

Throughout the book, we also provide exercises, activities and other tools that we hope will stimulate thinking and provide useful guidelines when seeking to develop creative ideas.

Whilst an understanding of the creative process isn't necessary to be creative—we've all got the capacity to be creative in one form or another regardless of our understanding of the process—such knowledge can certainly help maximise one's potential to produce creative outcomes, on an individual level, and within groups, teams, organisations, and societies.

The capacity to walk or run is innately human, though we must still develop the mechanical skill, and those wanting to run faster would do well to understand the mechanics of how they run and devise a training routine to improve this. Likewise, we can improve our innate capacity to be creative by understanding the processes in play, and then maximising our abilities in those areas.

The book explores creativity in the following way:

Chapter 1 explores the nature of creativity, suggests some definitions for the phenomenon and clarifies the differences between creativity, innovation and problem-solving.

Chapter 2 explores the creative person and discusses where the human capacity to be creative comes from. It also discusses the extent to which personality traits and skills have an impact on this capacity.

Chapter 3 examines the conditions for creativity to emerge and explores how they might be enhanced. What are the contextual elements that have an impact? What has been tried before?

Chapter 4 examines the barriers to creativity, and provides some tools and exercises to help overcome the roadblocks. We explore the role of persistence and perseverance, linked to a capacity called 'grit', and examine approaches to building resilience and mental toughness as support for the creative process.

Chapter 5 describes a selection of tools and idea generation techniques that can be used in a variety of settings, particularly in groups and organisations, to help come up with numerous diverse ideas, and to spark new insights and connections.

Chapter 6 revisits the cultural and contextual conditions that impact creative outcomes and examines how to create more innovative cultures. It explores a range of tools for stimulating innovation through improved organisational design, management and supervisory styles, rewards and resource allocation.

Chapter 7 examines processes for problem-solving. It also discusses approaches for facilitators who run creative sessions, as well as trainers seeking to inculcate some of the ideas in this book into their organisations.

Chapter 8 examines how to take good ideas into the real world and how to make them a reality. It also explores how to deal with the various hurdles that can arise in response.

Chapter 9 explores the impact of our creative cities and societies on our capacity to be creative. It examines some of the environmental factors in creativity and discusses societal approaches to creativity.

Chapter 1
Creativity, Innovation and Problem-Solving

What Is Creativity?

One of the most common 'creativity' tests is this: How many uses can you find for a paperclip in 5 minutes? Try it!

(See the Appendix for a list of possible ideas!)

Very simply, the more uses you can come up with, the more 'creative' you are deemed to be.

This test actually measures your capacity to think *divergently*—that is, to come up with multiple possibilities or solutions. It is the opposite of *convergent* thinking, which focusses on one fixed (and correct) solution. We rely on convergent thinking multiple times every day.

When we encounter an everyday problem, such as needing to commute to work, for example, we tend to rely on our memory of what has worked well in the past and repeat that action. However, resorting to tried and tested methods will not help us in our efforts to think creatively, because as we will see, a *novel* approach is one of the key ingredients of creative output.

Divergent thinking, therefore, is frequently part of the creative thinking process, because we need to envisage multiple possibilities that might lead to an original—and valuable—outcome.

Given how highly valued creativity has become in many areas of life today, there is an increasing desire to determine people's levels of creativity. Hiring managers at technology firms, for example, want to hire the most creative employees, and need to gauge the creativity levels of potential candidates so that they can do so effectively.

To measure creativity, however, we need to know what we're measuring. So, what is creativity?

Creativity, and the processes behind creativity, have attracted the attention of researchers from a range of disciplines—neuroscience, sociology, biology, organisational psychology, the arts—each with a focus on different aspects of creativity.

In an effort to pinpoint the essence of creativity, these researchers have measured various aspects of the creative process. From tests in divergent thinking like the one above, given to wide cohorts of subjects, to case study explorations of individuals famed for their 'creative genius', such as Albert Einstein or Steve Jobs, the creative process has been examined from a variety of different angles.

Neuroscientists, for example, have observed what happens in our brains as we experience creative thinking. In one such experiment, volunteers' brains were scanned as they solved riddles; in another, jazz pianists improvised on adapted keyboards whilst they lay in an MRI chamber.

Meanwhile, organisational psychologists have examined the dynamics in groups and teams that lead to the most creative ideas. They have analysed management styles, team dynamics and organisational cultures in an attempt to pinpoint the factors that are most effective in promoting creativity within teams and organisations.

A range of different theories has emerged accordingly, and along with it, a challenge to settle on one clearly defined and universally accepted definition of creativity. As we noted earlier, the notion of creativity covers so many different processes and outcomes that the complexity of this challenge is hardly surprising.

A Definition of Creativity

There are many takes on how to define creativity. Here is a sample of definitions from practitioners at the forefront of their creative fields:

Creativity is:[2]

---

[2] There are many definitions of creativity, from all walks of life. These are a selection and others can be found on quotation websites. E.g., brainyquotes.com

- The emergence of a novel, relational product, growing out of the uniqueness of the individual on the one hand, and the materials, events, people or circumstances of his life on the other—Carl Rogers (psychologist and writer)
- The occurrence of a composition which is both new and valuable—Henry Miller (writer)
- The ability to make new combinations of social worth—John Haefele (CEO and entrepreneur)
- A special class of problem-solving characterised by novelty—Newell, Simon and Shaw (team of logic theorists)
- Any thinking process in which original patterns are formed and expressed—H. Fox (scientist)
- The power to connect the seemingly unconnected—William Plomer (writer)
- Fluency, flexibility, originality and sometimes elaboration—E. Paul Torrance (educator, academic, creativity researcher)
- The process of bringing something new into being—Rollo May (writer and philosopher)
- Imagining familiar things in a new light, digging below the surface to find previously undetected patterns, and finding connections among unrelated phenomena—Roger von Oech (creativity consultant)
- The ability to use different modes of thought to generate new and dynamic ideas and solutions—Carnevale, Gainer, and Meltzer (business consultants)
- Bringing into existence something not there before—something new—an extension of the knowledge base—Roger Dennard (businessman)

These varied definitions reflect the many differences in creative output that we discussed in the introduction. However, there appear to be some universal commonalities that stand at the heart of efforts to increase creativity. In particular, the concept of 'newness' or novelty frequently appears.

Novelty

When brothers Wilbur and Orville Wright successfully flew their airplane, the *Wright Flyer*, for 12 seconds on 17 December 1903, they had introduced something new to the world: the first motor-powered airplane.

Creativity is the process of *creating* something, the act of bringing into existence something new: a new idea, a new product, a new way of doing things. It may also be a new way of combining existing products, processes or ideas, or producing the unexpected. Novelty and originality are seen as necessary components of creative output.

Keith Sawyer (1) described creativity, as it occurs on an individual level, as "a new mental combination that is expressed in the world". Here, we see an emphasis placed on novelty, and also on combination—of existing thoughts and ideas.

However, this idea of novelty or 'newness' is not without its complications. If you came up with a new idea, only to subsequently discover that someone else had already had that idea, were you being creative or not? You were certainly exercising your creative capacities, as that idea was new to you. You were using knowledge and expertise and combining existing ideas to come up with something you believed to be new and original, a process that is at the heart of creativity. If the idea was new to you, then—from an individualist perspective—this was creativity in action.

Other people who encounter this idea for the first time might also deem it to be creative. However, others might have seen the idea expressed before, and therefore might not consider it creative.

For example, a teenager may hear a new song by a new artist and marvel at how creative it sounds, combining diverse musical styles to sound like no other current artist; that teenager's parents, however, might not think it quite so creative, as it sounds just like a favourite artist of theirs from years earlier. Is Lady Gaga creative with her combination of edgy pop music, outlandish costumes and controversy? Or is she doing what Madonna did 30 years earlier?

From a socio-cultural perspective, an idea must be perceived as being novel by society at large, not just on an individual level, if it is to be viewed as creative. But one perspective on creativity takes the view that nothing created is actually intrinsically new.

Rather, something that appears to be new is merely the reformulation of existing elements into different forms and patterns. It is a 'remix' of these pre-existing elements, reconstituting or recombining ideas that have already been expressed. Arguably, this process of combination is the source of all creativity.

A common example of creativity is the invention of the motor car. (As we mentioned earlier, and contrary to popular opinion, Henry Ford didn't *invent* the motor car, he developed and manufactured an affordable assembly line version of it.)

The notion of a motor car may have appeared new in the late 19th and early 20th centuries, but could actually be viewed as simply the combination of various concepts and processes that had been around for some time already—the combination of a combustion engine (which already existed in various stages of development since the late 18th century) and a horse-drawn cart (which had been around for hundreds of years!). It was this *combination* of elements which was novel rather than the elements themselves—not so much a brand-new invention as a new combination of existing ideas.

Indeed, film-maker Kirby Ferguson created a documentary called *Everything is a Remix* (2) that explores this point. He cites successful figures from various fields and demonstrates how their 'creative' ideas were built from a series of pre-existing ideas.

He shows how film makers like George Lucas and Quentin Tarantino recreated scenes and vignettes from previous movies, sometimes with complete transparency. For example, the Tarantino movie *Kill Bill* placed heroine Uma Thurman in a yellow jumpsuit that is identical to the one worn by Bruce Lee in the 1978 movie *Game of Death*.

The process of combining different products and services that are often seemingly unrelated has been a major source of innovation. Examples include combining: petrol stations with grocery stores, phones with music and video players, breakfast cereals with snack bars.

The Beijing subway has experimented with a ticket machine that accepts plastic bottles as payment, incentivising recycling by combining it with public transportation. *Cirque du Soleil* combines the circus with the theatre. For British radio station, Classic FM, it was the process of making a new combination between an existing musical style and an unexpected musical audience, by applying a 'greatest hits' approach to playing classical music and targeting a non-classical audience in the process.

As the author Jonathan Lethem commented, "When people call something 'original', nine out of ten times, they just don't know the references, or the original sources involved" (3). And Keith Sawyer explains "All creativity includes elements of imitation and tradition; there's no such thing as a completely novel work" (4).

In the world of art and music, this process of recombination has caused issues with copyright law. If a new song builds on musical elements that have already been used elsewhere, at what point is the new song infringing on the copyright of a previously existing song (i.e., copying its ideas)? There is no clear-cut answer to this question, and there is an argument that states that if copyright protection is exercised too broadly and stringently, creative output in the arts will be undermined. We explore this issue further in Chapter 9.

In accepting that creative output builds upon and combines existing ideas in new ways, we can then recognise that the building blocks of creative output are all around us in everyday life. As we will discuss in Chapter 3, paying attention to these creative building blocks is one way we can feed our creative capacities.

Value

But is being 'new' enough to be deemed creative? When considering the process by which new ideas are formulated by an individual, then maybe it is. But when an idea is expressed in the world within a socio-cultural context, another important factor comes into play.

On Wednesday 16th October 2001, a cleaner arrived at the Eyestorm gallery in West London to clean up after a launch party for British artist Damien Hirst's new exhibition. The cleaner found the usual debris to be expected after a launch party—overflowing ashtrays, coffee cups and beer bottles strewn everywhere. So, he proceeded to clear everything up into bin-bags.

Unfortunately, some of the tables of ashtrays, cups and bottles had actually been carefully arranged by Hirst at the party the night before to create an impromptu new work of art. Given the value of Hirst's work to date, this new artwork would have carried significant value too. Yet it looked like trash (literally) and was treated as such.

A similar incident occurred at a gallery in Italy in 2015. How was the cleaner to know that this was art and not trash? And how would we look at it differently if it was installed in a cabinet on display in the gallery? Would we see it as art?

Or would we still see it as trash? Would we think of it as creative? Or just rubbish?

This illustrates another fundamental question in the assessment of creativeness—does the output have any value or not? The only thing separating this pile of trash from being a creative piece of art is someone deciding that it has value as a piece of art. Besides novelty or originality, a sense of value or usefulness or appropriateness is an important element in the prescription of creativity. Not only must something be original, but it must have value to be considered creative. And this sense of value is highly subjective.

Modern art may be one of the hardest and most subjective areas to assess for value; many other creative ideas can be assessed by more clear-cut and straightforward metrics than individual taste. For example, a solution to an everyday problem might be considered creative to the extent that it actually solves the problem (in a new way).

A new invention may be assessed on the extent to which it experiences commercial success (e.g., Apple computers). Another idea might be assessed by the extent to which it receives critical acclaim (e.g., a novel that wins the Booker Prize). How we ascribe value to a creative idea can vary hugely according to context.

Not only can an assessment of value be highly subjective, it can also change over time. The work of French impressionist painter Claude Monet was initially shunned when submitted to salons, but his paintings have grown to be some of the most valuable in existence. Vincent Van Gogh only sold a small handful of paintings in his lifetime, including *The Red Vineyard*, for 400 francs just before his death.

Contrast this with *Laboureur dans un champ*, which sold for $81.3 million in 2017. Or *Portrait of Dr. Gachet* which sold for $82.5 million in 1990 (the highest price paid for a work of art at the time), having originally been sold by Van Gogh's sister-in-law for 300 francs.

More recently, the British street artist Banksy set up an experiment in New York. Banksy's work sold at auction for hundreds of thousands of dollars, but he wanted to examine the nature of value in artwork. Amongst the other vendors of artwork, postcards and souvenirs outside Central Park, a man set up a stall with a series of canvases in the style of Banksy's stencil work. The canvases weren't ascribed to any artist—and were assumed to simply be cheap knock-off copies of Banksy motifs.

They were priced at $60. The man sat there by the stall all day, and sold a total of 8 canvases to 3 people, some negotiated at a discount, for a total of $420. The next day, Banksy released a video of the experiment, confirming that he had set it up. Accordingly, the value of the canvases, now deemed to be 'genuine, authentic' Banksy creations, shot up significantly!

Incidentally, 3 artists set up an identical stall one week later, with identical 'Banksy' canvases for sale. Despite the fact that they made it very clear that these canvases were *not* actually by Banksy himself and provided a 'Certificate of Inauthenticity' with each one, they sold out within an hour.

Ultimately, we all have individual responses to the ideas and creations that we see, and the extent to which we ascribe some positive value—from utility to aesthetics—will influence the extent to which we view them as creative.

Creativity

Creativity, therefore, can be defined as the generation of something—a product, an idea, a way of doing things—that appears to be expressed in a new way, and has value, usefulness, or appropriateness. Or, more simply, creativity is the process of creating something that is 'new' and 'valuable' (whatever those two terms are understood to mean).

Innovation

The term innovation is often used to denote the end result of creativity, and 'creative' and 'innovative' are often used interchangeably. Whilst they are similar concepts, there is more to innovation, however. Innovation is the process of refining creative ideas and *turning them into practical solutions*. It is the successful execution of a creative idea and its introduction to the world.

This might entail the introduction of new products or services to the marketplace, new processes within an organisation, how an organisation is designed and managed, how a brand is seen by the customer, and so on. Innovation introduces change, and it requires the innovator to convince people of the value of this change.

This change can be in the form of small incremental change (e.g., increased processing speeds in a new version of a computer) or larger, more disruptive and transformational change (e.g., the change to the taxi/car service business brought about by Uber).

And such change may come from a number of areas; the researcher Robert Sternberg, a professor of human development at Cornell University, explains that, for organisations, sources of innovative change come from employees, customers, competition, the public and partners such as suppliers (5).

In the Japanese tradition that permeated approaches to manufacturing and production (employed notably at Toyota, the car manufacturing giant), this approach of implementing incremental change is called *Kaizen*—change for better, or continuous improvement. This incremental change is then contrasted with *Kaikaku*—radical, transformational and frequently disruptive change.

The vast majority of innovations are simple small steps of incremental improvement (Kaizen), with periodic breakthroughs to new levels of approach (Kaikaku). It's these latter disruptive changes that hold the potential for game-changing, and are the drivers of innovation, because old assumptions and methods no longer hold.

As such, these disruptive changes are also most likely to be rejected or attacked, as they challenge and undermine all those who flourish in the old system or framework. The more disruptive the change, the more likely there is opposition to this change from the current practitioners in the field (e.g., huge opposition to Uber from taxi drivers).

In the athletics world, the history of the high jump is an interesting illustration of the combined nature of incremental and transformational change, as high jump records change and improve over the years. A number of different techniques were employed over time, each with the aim of jumping higher than previous methods had achieved; whilst these new methods were initially inferior to previous methods, they quickly superseded them and achieved higher jumps.

Incremental change occurs when the high jump record is increased using the same technique as before. Disruptive change occurs when the record is increased using an entirely different technique. The graph of records below, illustrated with the highest jump by each method over time, as the methods overlap, highlights this point (6).

High jump records for different techniques
(incremental and transformational change)

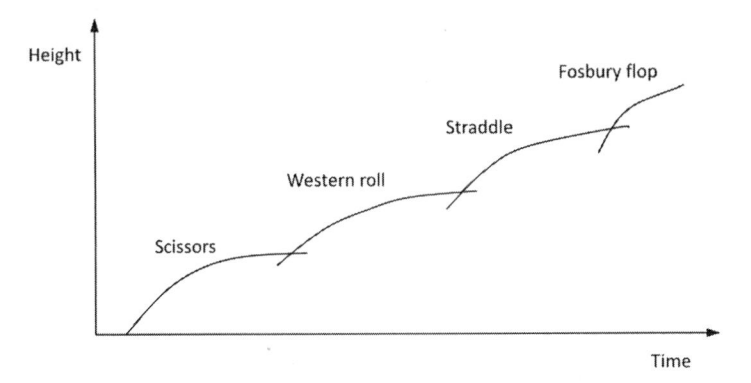

Several conclusions can be drawn from this graph:

1. If you want to win gold at the next Olympics, you won't do it with the 'scissors' method. Finding the next breakthrough is as imperative in sport as it is in commercial life.
2. A stimulus for seeking new (and disruptive) approaches arises when it becomes harder to squeeze incremental gains from existing methods. As the curve denoting improvement to the record flattened for each method (i.e., the increases in the record got smaller and smaller), people sought out a new method. This is particularly salient in a business setting: as growth from improving upon existing approaches gets smaller and smaller, new transformational approaches are required.

Gillette has been a leader in the shaving market for many years by continually improving their products. However, a new competitor called Harry's has made dramatic inroads into the shaving market by offering cheap quality products online through a regular subscription.

Gillette had failed to recognise that incremental change opportunities had become limited, and failed to innovate in its marketing approach, and so provided an opportunity for a new entrant to the market to bring disruptive change.

3. The early adopters of each new method did not jump as high as the experts in the old method (e.g., early adopters of the 'western roll' did not jump as high as the best jumpers using the 'scissors'). You can imagine that these innovators frequently heard the phrases, "You're wasting your time," and "If it ain't broke, don't fix it."

   Protection, encouragement and support is crucial if the new idea is to improve through persistence and not to die an early death from the discouragement of its users. Whilst it may appear that each method of jumping was clearly defined at the outset, the actual process of introducing a new method is usually much messier. Experimentation is a crucial part of the process, with sufficient time, resources and support required to do so.

4. These two processes of incremental improvement and transformational change should be equally pursued within organisations where possible. Apple undertakes a process of incremental change with the updates to its products. For the iPhone, there are numerous innovations that bring about incremental improvement with every new version of software and hardware—for example, a greater pixel count for the camera, an improved battery life, a bigger screen and lighter weight, etc.

   However, new versions often include transformational, disruptive change too—and it is frequently this disruptive change that will cause users to upgrade but can also cause outrage. One example was Apple's decision to remove the standard headphone jack from the iPhone 7 onwards. This corresponded with the growth of Bluetooth speakers and headphones, but also frustrated a lot of users who were comfortable with a standard headphone jack.

Creativity is an integral part of innovation, but innovation describes a wider process that accounts more fully for the successful execution and implementation of the creative ideas in a particular context.

Problem-Solving

This process is focussed on solving a specific problem and may rely on tried and tested approaches (using convergent thinking) or may require creative thinking to come up with a new approach to solving the problem. Its success is generally judged by the extent to which it solves the problem in question. Creativity may therefore play a role in problem-solving, but is not necessarily always involved.

In Chapter 7, we will describe a powerful approach called team-based problem-solving.

The 'Aha!' Moment – A Process Model of Creativity

The story of Archimedes that we cited in the introduction highlights a common characteristic of insight: the crucial insight came about after a period of time working on the problem, and then a period of time not thinking about it. After all the work Archimedes had put towards the problem, it was as he was relaxing in his bath that the insight came.

As such, we can often observe a linear set of stages in the creative process that include research and hard work, followed by impasse arising from an inability to make progress or from exhaustion, which in turn leads to relaxation, and finally to insight.

In 1926, English social psychologist and London School of Economics co-founder Graham Wallas, sixty-eight years old at the time, penned *The Art of Thought* (7). The book proposed an insightful theory outlining the creative process in four stages, based both on his own empirical observations and on the accounts of famous inventors and polymaths. The stages he outlined were:

1. Preparation
2. Incubation
3. Illumination
4. Verification

Preparation

This is the starting point of the process, brought about by a seed incident or catalyst. There may be a need to solve a specific problem that has arisen, or a dissatisfaction with the status quo. There may be a curiosity about the way things are done, or a dream of doing something that is not currently possible.

During the preparation phase, some initial research and experimentation may be undertaken to understand the scope of the issue in question, and potentially to increase one's knowledge and expertise in the field in which the issue resides. There then follows a period of further research, experimentation, and endeavour, usually accompanied by failure, frustration and eventual impasse.

Incubation

With such impasse, the issues may be put to one side, on the mental back-burner, and so begins the incubation period. During the incubation stage, the problem enters the unconscious mind, where synthesis of the various elements of the problem occurs. As we know from emerging neuroscience, the brain continues to work on the task subconsciously, and there is evidence that periods of rest and relaxation, and of putting the task to one side and engaging in another activity, can lead to insights emerging from the subconscious.

When we say "I'm sleeping on it" while trying to solve a problem, we are describing the incubation stage. And it may well be that this relaxation is entirely necessary to finding the answer—the harder you push, the less likely the answer is to arrive.

A review of experimental literature on incubation in problem-solving and creativity, conducted by Rebecca Dodds and her colleagues (8), revealed that 29 out of 39 experiments examined found a significant effect of incubation. The reviewers also suggested that incubation length and preparatory activities can influence the positive effects of incubation.

Paradoxically, therefore, the creativity process can sometimes be facilitated by deliberately not thinking about the problem but diverting the mind to something else, allowing the subconscious to work unguided.

Henri Poincaré, the great mathematician, and an influence on Wallas' writings, paid tribute to the unconscious self: "It is not purely automatic; it is capable of discernment...it knows how to choose, to divine...it knows better how to divine than the conscious self, since it succeeds where that has failed"

(9). And such is the unconscious nature of the incubation stage that we can never be sure where it will lead us.

In describing the creative process, Bill Watterson, creator of the Calvin and Hobbes cartoons, said, "The truth is, most of us discover where we are headed only when we arrive" (10).

The power of this incubation process has long been a mystery to the creators themselves. Einstein said, "The intellect has little to do on the road to discovery. There comes a leap in consciousness, call it intuition or what you will, and the solution comes to you, and you don't know how or why" (11).

The famed composer and conductor Aaron Copland once commented, "Inspiration may be a form of super-consciousness, or perhaps of sub-consciousness—I wouldn't know. But I am sure it is the antithesis of self-consciousness" (12).

With such obscurity around the insight process, therefore, it is tempting to place the onus on some mystical force in the universe to bring a flash of insight, and to treat the incubation phase as an exercise in sitting around waiting for inspiration to strike. However, there is only a chance of insight appearing if significant work or research has taken place first.

Knowledge and expertise in a particular field or domain are usually a prerequisite to insight and creativity, and the hard work of research and experimentation (and the inevitable accompanying failures) are a necessary prelude to the incubation phase.

As Nobel prize-winning chemist Louis Pasteur famously said, "Chance favours the prepared mind." We will discuss the role of this domain-specific knowledge or expertise in Chapter 2.

Incidentally, various cultures, communities, and civilisations have seen pause and rest as an important part of human centring—a way to integrate, and to 'let the soul catch up with the body'. The American poet Doug King tells the story of an explorer long ago who was making a forced march to the coast in order to catch a ship.

One day out from the coast, the porters sat down and would not go on, saying that they needed time for their souls to catch up. The explorer protested but they would not move. "Learn to pause or nothing worthwhile will catch up to you," concluded King, as a life lesson in learning to pause and be present (13).

This may be an apocryphal story invoking a spiritual explanation, but interesting research in the field of neuroscience is beginning to shed light on the powerful role of the subconscious, and the importance of balancing hard work and conscious exertion with relaxation.

Intimation

Imitation was not originally included in Wallas' steps in the creative process, but researcher Eugene Sadler-Smith (14) suggests that the addition of an 'Intimation' stage is a more authentic representation of Wallas' explanation of creativity. Intimation occurs when an insight is very close to arriving, and appears as an increasingly conscious train of associations or thoughts that lead to the 'aha!' moment of illumination.

Sadler-Smith's version of the model also includes three levels of *proximity to consciousness*: non-consciousness, fringe consciousness, and consciousness. Preparation, illumination and verification are all conscious processes, whilst incubation is non-conscious. Intimation is a fringe consciousness process, as the 'flash' of illumination begins to emerge from the incubation process.

It may then be that other activities need to be dropped and all focus put on the creative task at hand to bring the creative idea from intimation to illumination, where the insight occurs.

Illumination

This is the 'aha!' moment, the point at which the answer (or potential answer) arrives in the conscious mind. Thanks to the incubation process, the answer finally becomes apparent.

Sir William Hamilton's discovery, in 1843, of Quaternion theory, a mathematical solution to complex numbers as applied to mechanics in three-dimensional space, was preceded by 15 years of research and experimentation, along with frequent frustration and failure. Finally, he told his wife that he was quitting the problem and would waste no more time on it. But shortly after this, as he was crossing a bridge in Dublin with his mind finally free of the problem, a stroke of insight brought him the solution he had been searching for, and he carved the preliminary mathematical formulae on the parapet of the bridge.

Verification

The final stage in the process is verification, whereby the creative idea is tested out, to see whether it solves the problem, or has merit and value in some other capacity. This is where the practicalities and challenges of turning the idea into reality (and undertaking the process of innovation) begin to emerge.

Of course, there are many instances of creativity that do not involve flashes of inspiration and explicit 'aha!' moments that arise out of incubation. Instead, a creative outcome may be the result of a number of conscious experiments—Arne Dietrich's 'trial and error' model—or the result of choosing an option from a series of ideas from a brainstorm.

Nonetheless, this process model of creativity is useful when considering the role that incubation plays in the emergence of insights and 'aha!' moments, and it also highlights the power of the unconscious mind in our quests for creative ideas.

Chapter Summary and Conclusions

- Creativity, therefore, can be defined as the generation of something—a product, an idea, a way of doing things—that appears to be expressed in a new way, and has value, usefulness, or appropriateness.
- Creativity often involves divergent thinking, the process of considering many different possibilities or solutions.
- Creativity, innovation, and problem-solving are interlinked but distinct phenomena.
- Innovation brings about change that can either be incremental or transformative. Transformative change is often disruptive to some degree.
- In certain instances, the creative process undergoes the following stages:

  o Preparation
  o Incubation
  o Intimation
  o Illumination
  o Verification

Chapter 2
The Causes and Components of Individual Creativity

Introduction

Think of someone you believe to be creative. Not too hard a question, right? A few names of well-known individuals probably quickly spring to mind. People like Steve Jobs, Maya Angelou, Pablo Picasso, Beyoncé Knowles and Quentin Tarantino frequently appear as answers to this question. This isn't a particularly challenging question, but if we now asked you to tell us what you suspected to be the *cause* of that person's creativity, would it be so easy to answer?

Why is that person creative? What are the ingredients or behaviours that have led to that person's creative success? When the Wright brothers set about achieving the first airplane flight, what was it about them and their behaviour that led to this creative breakthrough?

When Marie Curie won her *second* Nobel prize, in 1911, for her insights and discoveries in radioactivity, what were the reasons for her sustained success in the lab? When the British band Radiohead composed *OK Computer*, an album frequently hailed as one of the best of all time, what were the ingredients to this creative triumph?

This is a much harder question to answer! The factors that lead to creative output are frequently misunderstood and often hidden from the casual observer, and as a result, they can be very hard to pinpoint. Furthermore, even these people themselves may not really understand what it is that has allowed them to exercise their creativity so impressively.

One of the most intriguing—and burning—questions on the subject of creativity is around this genesis of creative outcomes within individuals. Where does creativity actually come from? How and why does creativity appear on an individual level? Is it linked to specific personality traits or characteristics? Or is it linked to a set of behaviours, skills or processes? Are some people born creative whilst others aren't? Are some people inherently more creative than others? Or can anyone be creative if they want to be?

To put it another way, is our capacity to be creative based on *who we are* (our personalities) or *what we do* (our actions)?

On an individual level, the answer to this question may be critical for our careers and our livelihood. The business world often values creativity more than any other employee skill or characteristic, but do we actually have the capacity to increase our own chances of success? Can we do anything about it, or are we simply stuck with the creativity levels that we were born with?

Not surprisingly, this question has been the subject of much research since the mid-20th century, and various theories have emerged on either side: some theories focus on the *psychology of personality*, whereby certain personality traits are linked to creativity, and others focus on the *psychology of cognition*, arguing that creativity results from the use of certain thinking skills, behaviours, and capacities.

Unfortunately, definitive answers to the questions that arise about how and why we are creative remain elusive, but by exploring the ongoing debate and research, we are able to address the confusion and highlight some principles that will help.

Who I Am vs What I Do

One common approach to exploring the magic ingredients in creativity is to examine certain individuals who have enjoyed great creative success and are known for their creative output. By looking at the various traits, behaviours and characteristics of such 'creative' people, we might be able to observe certain key ingredients appearing time and time again, showing us the secret to creativity.

Who I Am

Across numerous studies into 'creative' individuals, certain personality-based characteristics are frequently found to be present, giving rise to various theories that our capacity to be creative is tied to the presence of these certain personality traits. Such reports tend to describe these people in terms of who they are rather than what they did, and attribute their creative success to these characteristics.

Simon van der Meer, a nuclear physicist who was dubbed 'the tamer of subatomic particles', yielded some impressive breakthroughs in particle physics throughout his career.

According to his obituary, the key to his creative success was his inherent great curiosity, combined with the fact that he was 'a natural tinkerer', evident in his fascination with creating household gadgets from scavenged parts when growing up in Nazi-occupied Holland.

Einstein's genius is often attributed to his 'headstrong personality', his fierce and impressive intellectual capacity, and his indefatigable curiosity. And Richard Branson's unconventionality—typified in the title of his how-to book *Screw Business as Usual*—is heralded as the key to his creative success.

Accordingly, one of the most commonly appearing personality traits in studies of 'creative' people is 'openness to experience', one of the five aspects of personality in the widely used five-factor model of personality (the other factors are: extraversion, conscientiousness, agreeableness and neuroticism).

'Openness to experience' broadly describes our comfort with and inclination towards new experiences, often accompanied by a curiosity to discover and investigate these new experiences. Since an important aspect of creativity is the novelty or originality of the output, it figures that this openness to experience could be important among creative people.

Educator Anna Craft highlighted a variety of characteristics that are frequently found in studies of creative people, and presented the following list of characteristics and traits that 'creative people' are frequently found to possess (1):

Openness to experience	Enjoyment of experimentation	Lack of feeling of being threatened
Independence	Sensitivity	Goal orientation
Self-confidence	Personal courage	Internal locus of control
Willingness to take risks	Flexibility	Originality
Sense of humour	Unconventionality	Self-reliance
Playfulness	Preference of complexity	Persistence

Further famous stories of creative triumph support these findings. The Italian engineer and architect Filippo Brunelleschi's most famous work was the dome of Florence cathedral. However, he bid for the contract without ever having built something as large before, and in the competition for the contract, he told the committee that they would just have to trust him to come up with a creative solution.[3]

Having won the contract, he then had to invent certain machines specifically for the project in order to complete the cathedral. In doing so, it was noted, he demonstrated considerable self-confidence and a willingness to take risks.

The frequent occurrence of such personality-based traits among 'creative' people has led to various theories that it is the presence of these personality traits that is key to being creative; that is, your capacity to be creative is a result of *who you are*.

But if creativity is tied to personality, then the implication is that you either *are* creative—and therefore capable of producing creative work—or you *aren't*, you're either born with it or you're not. Since your personality is set from a young age and will only marginally shift over time, the implication is that your creativity levels are similarly static.

This does not bode well for those of us who are considering how to increase or improve our capacity for creative output! As you read the list of traits above, you may have been thinking 'that's me' or 'that's not me'… and this unfortunately can easily lead us to defining ourselves categorically as 'creative' or 'not-creative'.

Some personality tests even include creativity as a specific measurable personality factor, based on the presence of some of the traits discussed above.

---

[3] In order to demonstrate his creativity and his capacity for unconventional solutions to the committee, he invited all the other participants to solve the problem of getting an egg to stand upright. They all failed. He then proceeded to crush the egg at one end and stood it up. When his opponents cried foul and claimed that he had broken the implied rule of maintaining an intact egg, he merely said that they could have done the same, and that they would also be able to make the dome if they were to see his design. The implication is that creative individuals are not bound by external rules or the implied limitations that our minds set.

A study by Thorne (2) using a test called the MBTI[4] creativity index showed that creative people are more likely to be extravert (as opposed to introvert), intuitive (as opposed to sensing), thinking (rather than feeling), and perceiving (rather than judging).

The Occupational Personality Questionnaire (OPQ) includes a personality factor called 'creativity and change', which is based on sub-factors of personality measures: unconventional, conceptual, innovative, variety-seeking, and adaptable. And another personality inventory, Cattell's 16 Personality Factors, identifies a second order factor called 'creativity' that is a construct based on nearly all of the primary factors.

All these tests reinforce the idea that an individual's capacity to be creative is a fixed aspect of personality, and that people are therefore reliably more or less creative.

The Problem with Personality Tests and Traits

However, despite a number of traits and characteristics appearing frequently in the study of people who have achieved great creative output, there still exists no personality test that can consistently predict creativity based simply on the presence of certain personality traits.

If creativity was caused explicitly by the presence of certain traits in people, then we should be able to predict that people who have these traits will be more creative than those without them.

Put simply, people scoring highly in the pertinent traits such as openness to experience when administered a personality test should also score more highly than other people in other tests of creativity. But this is not reliably the case. The presence of these personality traits does not guarantee an increased capacity to be creative. They may be helpful in efforts to be creative, but are not solely responsible for this creativity.

Albert Shapero, a professor at the Harvard Business School, once commented, "Despite several decades of research effort on creativity and highly creative individuals, there is as yet no profile or test that reliably predicts who will be highly creative in the future" (3).

---

[4] The Myers Briggs Type Indicator (MBTI) is a popular personality diagnostic tool, measuring personality across four dimensions—extravert/introvert, intuitive/sensing, thinking/feeling, perceiving/judging.

If creativity is not solely linked to our inherent traits—something we have in a consistently fixed individual level—then where *does* it come from?

What I Do

In some fascinating research, author Mason Currey (4) explored the daily rituals and routines of numerous renowned creatives, gathering together their reflections on their creative processes to uncover some of the tricks and techniques they employed to help them with their creative work.

From long daily walks to precisely timed naps, the creatives employed a number of approaches to their creative work to try and get the best out of themselves. Japanese author Haruki Murakami does all his creative writing in the morning, rising early and working through to the middle of the day.

The rest of his day is given to leisure time and exercise (he is a passionate long-distance runner). Picasso, on the other hand, did not undertake his creative work until the afternoon, and worked late into the night.

The Russian composer Tchaikovsky took a short walk every morning before he started work, and then another walk after lunch, frequently stopping to make a note of ideas that struck him as he walked. Fellow composers Beethoven, Mahler and Benjamin Britten also relied on a daily walk to get their creative juices flowing. Somerset Maugham liked to compose the opening sentences of his day's writing while taking his morning bath, while poet W.H. Auden kickstarted his day with a hit of Benzedrine, a brand of amphetamine!

Many of these people attributed the success of their creative capacities to these elements of their routines, and it can be tempting to think that if we mimic their approaches, we too can boost our creative capacities. While there may be useful approaches to be gleaned from the rituals of creative people (and research shows that walking might indeed stimulate our capacity to think creatively), we must remember that these behaviours alone are not the answer.

They are not necessarily the ingredients for creativity that are universally applicable to us all so much as the factors that helped stimulate the creative faculties of *that particular individual.* What works for one person in one specific time or context doesn't necessarily work for everyone. We will explore the factors that might work for us individually in Chapter 3.

However, while many of these habits tend to be idiosyncratic, there are patterns that start to emerge in the behaviours that creative people display. In another review of studies of people deemed to be creative, journalist Carolyn Gregoire (5), amongst others, identified a number of common characteristics:

- They work the hours that work for them, though invariably they work hard.
- They take time for solitude.
- They turn life's obstacles around.
- They seek out new experiences; e.g., take a different route to work, read something they wouldn't normally, visit a museum/art gallery/industrial workshop—something different to the normal pattern of their lives and interests.
- They fail constructively—they see failure as a learning opportunity, they ask themselves what they can take from the failure in order to do something better (or at least differently) next time.
- They ask big questions—they start with the end goal and ideal outcomes in mind.
- They people watch—they notice the people they encounter, their mannerisms, clothes, actions, etc.—and observe the environment around them.
- They take risks; e.g., being the dissenting voice in a meeting, or investing time (or money) that could be lost.
- They view life as an opportunity for self-expression—they treat life as a journey of discovery of the world and of themselves.
- They follow their true passions.
- They see new possibilities.
- They follow their intuitions.
- They get out of their heads and look at things through many different perspectives. A study by Dyer and Gregerson (6) found that innovators spend 50% more time than others trying to think differently in new ways and make new connections.
- They lose track of time; they daydream, which can facilitate 'creative incubation'; they allow themselves flights of fancy. As Albert Einstein once remarked, "When I examine myself and my methods of thought, I come to the conclusion that the gift of fantasy has meant more to me than my talent for absorbing positive knowledge" (7).

- They surround themselves with beauty—they arrange workspaces with pictures and objects that are stimulating, interesting and beautiful.
- They connect the dots, they look for patterns, they make connections (they use spider diagrams and mind maps—see Chapter 5).
- They constantly shake things up.
- They make time for mindfulness.
- They are sensitive to others.
- They think differently.

Unlike Craft's list earlier, this list includes factors that tend to be general behaviours and dispositions rather than fixed abilities, characteristics and traits. As such, anyone can adopt these tendencies in their own way, with a bit of practice over time, and harness them to fuel their creative capacities. By attempting to incorporate more of these behaviours and dispositions into our lives, therefore, we should have greater success at generating creative ideas. The challenge, then, is to ascertain how to actually make these approaches a reality, and we will discuss this in detail in Chapters 3, 4 and 5.

However, we must recognise that simply doing all the things discussed above won't *necessarily* make us more creative. Rather than a specific blueprint set of actions that lead to creative output, all these observed behaviours and dispositions are the manifestation of some deeper principles at work in our creative endeavours.

What is most important is to understand *why* these behaviours and dispositions listed above can help, and what the underlying principles behind them might be.

When we understand these underlying principles that fuel creativity, we can start to figure out how to approach our creative work on our own terms. We can then decide how best to incorporate these principles into our own lives and creative practices, and how to develop strategies that will work best for us in the challenges we face.

A Caveat for Research on Creative Individuals

Before we explore these underlying principles further, we want to highlight a few general caveats that arise from the study of 'creative people'.

Creativity is most often domain-specific, usually in the areas in which we have great knowledge or expertise. Just because Quentin Tarantino displays great creativity in his screenwriting does not mean that he is necessarily creative in other domains too (e.g., engineering or marketing). Accordingly, we are all capable of 'every-day' creativity because we are all experts in our own lives.

As we mentioned above, what works for one person in one specific time or context doesn't necessarily work for everyone. There are many ways to facilitate the types of thinking that are present in creative output, as we'll discuss when we talk about idea generation techniques and other practices that facilitate creative work in Chapters 3, 4 and 5.

Some techniques will be better suited to certain individuals than others, some will also be better suited to certain situations than others. How Quentin Tarantino behaves or conducts himself—or displays certain personality characteristics—when he is writing the script for a future blockbuster or masterpiece will not necessarily work for any other screenwriter in Hollywood.

The behaviours, traits, and dispositions that we can see in action during the creative process aren't the whole story. What about all the previous work Tarantino has done to hone his scriptwriting and build expertise? Or all the movies he's watched in the past that have inspired ideas waiting to be included in a new script?

It can be easy to ignore the less time-specific factors, such as prior knowledge and expertise, or persistence and perseverance in experimentation (and failure) over time. We also tend to ignore any situational, contextual, and domain specific influences too, such as simply being in the right place at the right time.

Finally, these people are not *always* 'creative'. In so far as creative ideas introduced to the world are deemed creative by virtue of what an audience thinks, or by what commercial success they might enjoy, no creative practitioner has a constant success rate (although some groups and organisations have come close, for example, Pixar and their string of film successes).

The inventor Ron Hickman was responsible for creating the Workmate, a DIY workbench that was a huge bestseller for Black and Decker. However, despite many attempts, he was never able to invent another product that had similar success.

The Underlying Principles of Creativity

When we broaden our observations beyond those specific people deemed to be 'creative' and examine the factors that are present when any individual is using their creative faculties—in organisations and workplaces, in university classrooms, in their day-to-day lives—some underlying principles become apparent.

Creativity usually transpires in a subject area that the creator knows a lot about. This could be jazz or quantum physics or marketing; whatever the domain, knowledge or expertise is present. We see people use certain modes of thinking, asking questions such as 'what if this were possible?', exploring lots of options, and using their imagination to conjure up novel ideas or to consider what might happen if certain existing ideas were combined. And finally, we see a level of motivation, persistence and determination to get the work done, overcome self-doubt and procrastination, and see a particular project through to its conclusion.

Teresa Amabile (8), an academic from Harvard who has spent considerable time researching creativity and working with organisations to improve their creative output, identifies three elements at the heart of these observations.

According to Amabile, the key ingredients to the creative process—and the underlying principles that drive the pertinent behaviours and dispositions we discussed above—are: expertise, creative thinking skills, and motivation. Tina Seelig (9), who teaches innovation and creativity at Stanford, encapsulates the three important ingredients in a similar way: knowledge, imagination, and attitude.

Let's examine them in more detail now.

Knowledge and Expertise

Whatever creativity is, it is certainly a complex process containing a multitude of elements. A deep level of knowledge and expertise or mastery in a particular domain are often the starting point for creative work. Why would you need knowledge if creativity is all about coming up with novel ideas?

Well, creative outcomes invariably take existing ideas and reconstitute them, combine them with other ideas, re-arrange them, or find new contexts in which to express them. Making unfamiliar connections of familiar ideas requires you to be familiar with those ideas in the first place.

We have already looked at several examples in which lengthy study or research has been the bedrock of subsequent insight, and people viewed as 'creative geniuses' often have a high level of expertise in their craft, the domain in which they are viewed as creative geniuses. As a child prodigy, Mozart had already acquired extraordinary expertise in music and this expertise underpinned his compositions.

But even for less exceptionally accomplished people (i.e., the rest of us non-geniuses), creative output often builds on a level of expertise in a particular subject—whether a specific academic discipline, business market, hobby, or simply our own everyday lives—that is developed from a lengthy period of knowledge accumulation and practical experience.

When insight does suddenly strike, seemingly out of nowhere, it is the accumulation of all this knowledge and expertise that paves the way for this to happen. When we focus on what was going on at the point of insight, we tend to forget about all the hours of building knowledge and expertise that made this moment possible.

Some knowledge and expertise are highly domain specific, other knowledge and expertise are more broadly applicable to the human condition. For most of us, it would be extremely unlikely that we could come up with a creative approach to quantum physics because we know very little about it.

However, we are much more likely to have the capacity to come up with creative marketing ideas, because in our own ways, we are all experts on marketing—we're all consumers who are marketed to, so we have plenty of experience of the marketing process.

One advantage of creative teams over creative individuals is that teams can bring together the many forms of expertise held by those teams' members that are needed to solve large or complex problems. As Nobel laureate, economist and psychologist Herb Simon noted, "expertise creates a network of possible wanderings" (10), in the intellectual space used to explore and solve problems. The larger this space the better.

Expertise includes technical, procedural and intellectual knowledge of the subject at hand, as well as skills of application. This is the know-how that often takes individuals years to accumulate, and frequently includes many hours of practice—the 10,000 hours that Malcolm Gladwell famously described (11)—driven by curiosity, passion, or personal interest. It provides a vast resource of appropriate information and ideas to be combined in new 'creative' ways. An accomplished painter is drawing upon the development of many skills: acute observation of light and shade and the subject itself, technical painting skills in the application of different media and materials, and a wealth of experience in composition.

For a filmmaker, it might be a deep knowledge about the theories of filmmaking, technical knowledge about equipment and editing techniques, and a rich knowledge of the history of film, as well as many hours logged watching countless movies.

This knowledge provides the raw materials with which to be creative. Ideas can be combined, reconstituted and developed in new ways to provide creative output. A number of great jazz musicians who excelled in their improvisational playing, were also highly trained in a formal musical setting.

The Juilliard School, a prestigious school of music in New York City, boasts a number of successful jazz musicians as alumni, including Chick Corea, Nina Simone, and Miles Davis (although the latter only lasted one semester). These highly trained musicians knew the playing field and they knew how to experiment with it, reconstitute it and subvert it as a result.

In the movie *Shine*, the Australian pianist David Helfgott (played by Geoffrey Rush) wants to play Rachmaninov's *Piano Concerto No. 3*, which is extremely challenging. The director of music (played by John Gielgud) tells him that if he wants to perform this particular piece, he'll have to study the music in great detail, and then 'throw away the music and play from the heart'. Once we have developed a level of expertise, then we can play with the content of that expertise and knowledge to create, to express ourselves and to go *beyond* existing boundaries and constraints.

However, there are several caveats to bear in mind when it comes to the role of knowledge and expertise in the creative process.

Firstly, expertise can 'anchor' thinking to tried and tested methods and frameworks of the past, or to conventional wisdom or 'the way it is', and can lead to familiar patterns of thinking and problem-solving. This then limits the originality and diversity of the solutions or ideas that we generate (12).

People without significant knowledge of a particular subject area may be freer and wider reaching in their thinking, leading to more original solutions. However, these solutions are likely to be viewed as less appropriate (appropriateness being one of the two components of creative output, as we discussed in Chapter 1).

This provides an argument for combining top experts with relative novices within team-based situations. In a television commercial for a popular brand of taco shell, two groups of adults are seen arguing with each other in favour of either hard-shell tacos or soft-shell tacos.

As adults, they are anchored—from a lifetime's experience—of having to choose between these two options when making dinner choices. It takes a young child (the non-expert) to ask the question "Why not both?" ("¿por que no los dos?") to provide the creative solution—packs of taco shells that contain both hard-shell tacos and soft-shell tacos—thus keeping everyone happy!

There is also the risk that experts can become complacent in their expertise and ignore or fail to spot developments in knowledge and research in their field. We must therefore hold on to our expertise lightly, and be prepared to continue learning.

Furthermore, the trend for several hundred years, commensurate with the explosion in knowledge across that period, has been for greater specialisation in particular disciplines or subject areas, leading to deeper knowledge in these distinct areas. Whilst this depth of knowledge may well be beneficial to the creative process, a downside of specialisation is that it comes at the expense of a breadth of knowledge across disciplines.

This may therefore limit the cross-fertilisation of ideas across disciplines that can be integral to creative outcomes, and was important to the polymaths of the past. Leonardo da Vinci applied knowledge and skills from several disciplines—medicine, warfare, painting—to generate his creations, from paintings such as the *Mona Lisa* or *The Last Supper* to his sketched designs for a helicopter. A balance between depth and breadth of knowledge must be considered. In team situations, therefore, it can be useful to put together teams of people that have complementary bodies of knowledge rather than those who have identical knowledge.

Finally, ambiguity, paradox and gaps in knowledge are not a cause of despair for experts but an opportunity for creativity. As he was wrestling with atomic theory and Heisenberg's principle of indeterminacy, the nuclear physicist Niels Bohr once said, "How wonderful that we have met with a paradox. Now we have a hope of making some progress" (13). Arriving at the paradox actually clarified the issue to resolve, and being faced with a specific conundrum engaged the brain to find a solution.

Creative Thinking Skills and Imagination

Close your eyes... Imagine that it's 10 years in the future. What does the world look like? What does your life look like: what happens in your day-to-day? Who are the people around you? What does your house or apartment look like? And who is the president of the US?

In undertaking this exercise, you've just used your imagination to envision a future that doesn't yet exist. How we use our thinking processes, and the extent to which we can use our imagination to visualise something as yet non-existent or something that we have not yet experienced, is very significant in the pursuit of creativity.

We need to use our cognitive faculties in such a way as to conceptualise new ideas and new approaches, to call upon our existing knowledge, and combine, transform, or reconstitute it. So being mindful and deliberate in the way that you apply your thinking skills to the creative process is extremely useful—for example, by asking big questions, looking at things through many different perspectives, and so on.

In particular, a mode of thinking called *divergent* thinking is important. As we mentioned in Chapter 1, this is the process of exploring possibilities and generating many potential ideas, answers or outcomes. When we encounter an unexpected road-closure on our commute to work, we can use divergent thinking to start considering all the other possible ways we could get to work. To generate creative outcomes, it is important to employ divergent thinking rather than the *convergent* thinking style we use much more habitually.

Convergent thinking aims to find the one 'right' answer to the problem. It draws therefore on tried and tested ideas and familiar approaches to the many daily tasks we face, or on the application of standardised techniques, and it follows rules, norms and guidelines.

It is based on familiarity with the problem and the 'best' solution to it, whether through our own experience or the experience or opinions of others from which we can learn, sometimes with no further scrutiny or questioning. This is the route to work that we take every day, day in and day out (road closures not withstanding!).

We follow many habitual routines during the course of a day—our morning routine when we wake up, our commute to work, our evening meal preparation—and doing so relies heavily on convergent thinking. It would be exhausting and paralysing if we had to figure everything out from scratch every day, if we lay in bed every morning just after waking up and thought 'what shall I do now?'. We are creatures who find comfort in certainty, and convergent thinking provides this.

However, convergent thinking can be counter-productive when creativity is required. The mindset of convergent thinking is the enemy of creative thought because it focusses on finding the one correct or best answer, and this notion of what is correct or best is usually based on what has worked before.

Creative thought, on the other hand, needs to be free to wade into uncharted territory, to experiment with novel and untested ideas that may or may not work. This is where divergent thinking becomes useful.

Divergent thinking entails the generation of multiple solutions to any given problem or situation. It draws on ideas from all areas of thinking, and often involves merging, or combining, unusual ideas or unrelated concepts. There is no 'right' answer per se, and a key part of the process is generating (and potentially road testing) many different ideas, before then choosing some of these ideas to develop further. As we'll see when we discuss idea generation techniques in Chapter 5, aiming to generate a large quantity of ideas is a key part of the creative thinking process.

Divergent thinking is sometimes used as a synonym for creative thought, and divergent thinking tests (e.g., multiple uses for a paper clip) are often used as proxies for measuring 'creativity'.

As an example, consider approaches we might take if we needed to raise awareness for the release of some new music. Applying convergent thinking patterns to the problem might lead us to promote the new album through conventional channels, such as social media announcements and sponsored content, a media publicity campaign, incentives to streaming services and retailers, etc.

It's the kind of approach that has been employed by countless artists before, sometimes with considerable success, but it is certainly not a creative approach, and may not maximise the success that might be possible. Applying divergent thinking patterns, however, might lead to all sorts of ideas about publicity stunts, unusual promotional targets and various other measures for increasing awareness.

It was the divergent thinking inspired by a 'what if' conversation that led French electronic artist Daft Punk to stage the global launch of their album *Random Access Memories* at a country fair in rural Australia, as part of a campaign that generated considerable global awareness and excitement for the release of the album. Divergent thinking can be very productive in a situation like this where there really is no right answer.

So, what thinking processes would creative thinking skills include? Some of the thinking skills and capacities that are of importance in the creative process include:

- Using divergent thinking to generate a number of ideas
- Questioning assumptions—and the status quo—and asking why?[5]
- Asking 'what if' questions
- Accessing your imagination to envision novel scenarios
- Looking for ways to put existing ideas together in new combinations to form new concepts
- Exhibiting the capacity to find new uses for existing ideas
- Choosing good ideas to explore further

Many idea generation techniques aim to encourage divergent thinking, as well as upholding certain other principles through the process, such as generating a large number of ideas, and withholding judgement during the process of generating these ideas. These techniques help us to employ many of the approaches to thinking listed above. Chapter 5 lists a number of these techniques.

Otto Rank, an Austrian psychoanalyst and close colleague of Freud, stated that creativity is an "assumptions-breaking process" (14). This echoes Picasso's comment that creation involves destruction. Creative ideas are often generated when one discards preconceived assumptions and attempts a new approach or method that might seem unthinkable to others. Creativity involves developing an idea that would not have been expected in our usual behaviour, or usual thinking.

Challenging assumptions is often a key part of the process of creative thinking. We accept many practices and processes in life without ever questioning whether there might be alternatives.

Creative thinking, then, is the mental process of discovering "new ideas or concepts, or new associations of the existing ideas or concepts, fuelled by the

---

[5] There is a technique known as the '5 whys' that entails asking 'why' to every statement about an issue until you reach the root of the issue. We describe this technique in Chapter 5.

process of either conscious or unconscious insight". According to creativity consultant Roger van Oech (15), "Creative thinking involves imagining familiar things in a new light, digging below the surface to find previously undetected patterns, and finding connections among unrelated phenomena." Creative thinking refers to *how* people approach problems and solutions—their capacity to put existing ideas together in new combinations (16).

Intrinsic Motivation and the Attitude One Brings to the Process

In 1707, off islands to the southwest of England, the greatest disaster in British Naval history occurred. With no means of judging longitude, the British fleet was off course and many ships were lost when they hit the rocks at night. As a result, the British Parliament offered a large financial reward for whosoever could present a solution to the problem.

As a carpenter, John Harrison had become fascinated with building and repairing clocks, and in 1713, he set out to invent an accurate chronometer that kept perfect time at sea. It took many years and eventually, he produced four successful chronometers, each better than the last. He was motivated by a love of clocks and mechanical movements and the challenge of being the first person to solve the problem. (Though the prospect of a large prize probably helped as well!)

Charles Darwin's theory of natural selection came to him over several years as his knowledge from the voyage of the Beagle as well as his continuing observations of doves slowly formulated into his theory of natural selection. His curiosity, his passion for the subject matter, and his willingness to consider radical new theories motivated him to continue.

The attitudes that we bring to the creative process are instrumental in ensuring that the thinking techniques that we apply to the knowledge and expertise we possess result in creative outcomes. With the wrong attitude in place, our creativity is unlikely to bear fruit.

A significant factor in our attitude towards our creative endeavours is motivation. Without motivation, we are unlikely to persevere to find a creative solution. According to Teresa Amabile: "In our creativity research, my students, colleagues, and I have found so much evidence in favour of intrinsic motivation that we have articulated what we call the *Intrinsic Motivation Principle of Creativity*: people will be most creative when they feel motivated primarily by the interest, satisfaction, and challenge of the work itself—and not by external pressures" (17).

As we discussed in the introduction, much creative endeavour is not simply the result of a flash of inspiration out of nowhere, but is preceded by a lot of hard work, experimentation, trial and error, and failure. It takes a significant amount of motivation to continue with the process and not give up.

Thomas Edison allegedly tried over 600 materials as an element for his light bulb (though he never actually counted). He also held 1,093 patents in total. Beethoven wrote 32 piano sonatas, of which his famous *Moonlight Sonata* was just one. Picasso created some 20,000 works.

None of these creators arrived at the solution or the successful creative work just like that. They arrived there because of all the hours of experimentation and prior work that they'd undertaken, and their personal motivation to keep moving forward. As Edison himself famously quipped, "Many of life's failures are people who did not realise how close they were to success when they gave up." (18)

Nobel prize-winning physicist Arthur Schawlow comments, "The labour-of-love aspect is important. The most successful scientists often are not the most talented, but the ones who are just impelled by curiosity. They've got to know what the answer is." Creative output occurs because the creators are motivated to put in the time and effort to see the process through.

It is important to note the difference between *intrinsic* motivation and *extrinsic* motivation here. The latter is produced by external reward or punishment (carrots and sticks)—at work for example, we might be motivated to work hard by the reward of bonuses or promotions, or the threats of being fired or publicly humiliated in a team meeting.

Intrinsic motivation is internally generated—that is, the motivation comes from an internal belief in the inherent value of carrying out the action, or an enjoyment of doing it. It is *intrinsic* motivation that is key to the creative process: a passion for the subject and a drive, regardless of the reward, to find the answer to the problem being explored.

Einstein talked of the enjoyment of 'seeing and searching'. This can pose something of a paradox for companies that are seeking to accelerate the creation of new products and services by specifically tasking individuals and teams to be 'creative': how can they help to instil an intrinsic motivation in their employees rather than just providing external motivations? We will explore this issue in Chapter 6.

Another key aspect of the attitude towards the process that is important is dealing with all the derailers—self-defeating thoughts and behaviours—that we bring upon ourselves in the creative process: procrastination, self-doubt, fear of failure, laziness, etc.

To some extent, the challenge is not so much to *be* creative, but to *allow* ourselves to be creative and to harness our natural creativity by clearing these derailers out of the way. In Chapter 4, we will discuss a number of approaches to defeating these derailers.

Arguably, this is the most important aspect of the whole process, particularly when it comes to the process of innovation. All the good ideas in the world count for nothing if you don't have the perseverance to put them into action.

Some important research by Angela Lee Duckworth (19) and others has argued that a construct called *grit* is the best predictor for success in life. Grit is defined as having passion and perseverance for long term goals. And it's this quality that is necessary to see creative outcomes through. Grit can be increased in a number of ways. Duckworth's research shows that grit is a better predictor of success than IQ, family background, social intelligence, EQ, health, and many other measures traditionally taken as predictors of success, and we will discuss it further in Chapter 4.

Flow is another psychological construct that can play a role in the creative process. Hungarian psychologist Mihaly Csikszentmihalyi (20) has conducted significant research into the flow state and shown how it is often present in creative processes. Flow is described as a state of optimal experience in which:

- A person's skill levels match the levels of challenge that they face (and their skills are stretched)
- They are focussed intensely on the present moment
- They lose any sense of reflective self-consciousness, and their sense of awareness merges with their actions
- They feel in control of their actions, and gain a sense of effortless mastery
- Time appears to pass quickly
- They find the flow activity intrinsically rewarding

Flow can often be felt when improvising on a musical instrument, for example, or when skiing down a challenging run.

As well as the correlations with creative output, research on flow has shown positive correlations with enhanced well-being, as well as reduced stress and anxiety, increased productivity and innovation at work, greater engagement in life, and increased self-esteem. It also serves as a buffer against adversity and prevents pathology.

Chapter Summary and Conclusions

This chapter has suggested that there are behaviours that will have a positive impact on our individual capacity to be creative, whatever our personality traits. These are the behaviours therefore that should be developed.

In particular, there are 3 ingredients to the creative process that we should consider:

- Knowledge and expertise
- Creative thinking skills and imagination
- Intrinsic motivation, and the attitude we bring to the creative process

As Keith Sawyer explains (21), "You can be just as creative as anyone else. What makes someone creative isn't a personality trait; it's:

- mastering a creative domain
- the right work habits (for example, hard work interspersed with time off)
- knowing how to select good ideas
- knowing how to combine and connect your ideas
- based in the same mental processes that every human holds."

Chapter 3
Enhancing Our Individual Creativity

Introduction

This chapter aims to build on the two previous chapters and look deeper at how we can develop our capacity for creativity. Whilst it is unlikely that we might become a genius overnight, there are many things that we can do to help develop our creative faculties.

In Chapter 2, we discussed some of the characteristics that are commonly associated with creative people, as well as the ingredients of knowledge and expertise, creative thinking skills and imagination, intrinsic motivation and an appropriate attitude. Now in Chapter 3, we will look at a range of other factors that can influence our capacity to be creative.

This chapter aims to discuss several issues to consider in order to put ourselves in the circumstances most conducive to generating creative outcomes. We will discuss the following topics:

1. Personal conditions for creativity
2. Observing the world around us
3. The search for new ideas
4. Learning styles

The Personal Conditions for Creativity

The singer Adele once admitted to a particular approach in her song-writing process: "I'd get drunk and end up admitting things to myself in my diary." (1)

It might have been a flippant comment, but it revealed something significant about her approach to song-writing—that a level of emotional honesty was important in her song-writing, and a key part of her creative process, therefore, was to find a way to access this emotional honesty. (In this example, getting drunk and writing in a diary!)

Noel Gallagher, songwriter and guitarist for British rock band Oasis, explained: "The best time to write for me is when I first open my eyes in the morning. I've written some of my best things first thing in the morning—there's all these things going through your mind. *Champagne Supernova* started like that. In the morning you've had nothing to eat, you're a bit sleepy, you haven't watched the telly and you've had no time to be influenced by anything. It just comes straight off. You think about what you were dreaming about. I think *Some Might Say* was written very early in the morning." (2)

In an interview with *Time Out* magazine, playwright and screenwriter Rita Kalnejais reflected on her creative practices for a column entitled *How I Write*. She has developed a number of idiosyncratic practices to help her to get her creative juices flowing and to write. She finds that she works best early in the morning, so gets up before dawn, but also writes before she goes to sleep, "So my dreams can take care of questions I can't answer in the play." (3)

She covers her walls with notes on post-its, and writes ideas, notes, phrases on her desk, which is made out of an old door. She paints her nails to keep her engaged and alert as she types on her laptop, and writes ideas all over her arms and legs so that she's constantly reminded of them throughout her day. She gives herself regular dance breaks during her writing. As she admits, it's "very chaotic, but it works for me".

The people in these examples have ascertained the behaviours, habits, and practices that work for them and support them in their efforts to be creative and produce creative work. They can then replicate these features and make them part of their habitual process. And although these people are all in creative professions—the arts—and have therefore given some consideration to their creative process, there's no reason why such reflection can't be beneficial for any of us too.

As we are all individual and unique, we tend to be at our most creative under different conditions, conditions that are unique to us. Being aware of the situational context in which we are at our most creative is very valuable, as we are then able to recreate this context to help us be more successful in our efforts to be creative. It can be worth experimenting with various elements of the situation to see what works (and what doesn't).

So, what works for you? Think of the times when you've been at your most creative, when you've come up with a new idea for something or discovered a solution to a problem you're facing. What were the conditions in which you found yourself? What were some of the aspects of your situation—the environment, the time of day, the headspace you were in, or company you were with—that might have contributed to this creative output? Do you notice any patterns across the different occasions?

If you were trying to create the conditions to give you the best possible chance to be creative, therefore, what would those conditions be?

This situational context is your 'personal creativity profile'. Consider all the elements that contribute to your personal creativity profile. Write them down and refer to this profile when you want to be creative. This list is your personal set of guidelines for creativity.

Note that it's important to be aware of what part of the creative process you are wrestling with. Are you trying to come up with new ideas? Or are you trying to put in the hard work to see those ideas to fruition (e.g., writing the first draft of a novel once you've already sketched out the plot). Whichever part of the creative process you are working on will have a bearing on the conditions that are most beneficial.

Consider the following elements.

Time of Day

Are you a morning person or a night owl? Do you find it easier and more productive to work on creative projects early in the day or later on?

Author Steven King noted that he preferred to write new content in the morning, but then to focus on editing that content in the afternoon (4). Maya Angelou also preferred to write in the morning. As we described in Chapter 2, many creators and artists established daily routines in which they undertook their creative work at specific times of the day.

Consider how you can build a daily routine that allows you to access the best times of the day to be creative.

Energy Cycles

When are you at your most energetic? (And least energetic?) People have individual energy patterns and find themselves to be more energetic at different times of the day—perhaps peaking in their energy levels in the mid-late morning, and then dipping precipitously after lunch.

It can be beneficial to leave certain creative endeavours—such as those that require perseverance, or high cognitive activity—to times when you have high energy. Use low energy times for things you can do on autopilot.

However, some research suggests that being in a state of tiredness or low energy can jog you out of habitual thinking patterns and bring about more effective divergent thinking and unusual ideas (and therefore creative outcomes), particularly when the ideas or solutions require insight rather than analysis. As we saw earlier, this sleepy state—albeit at the start of the day—worked well for Noel Gallagher of Oasis. Other research has showed a similar effect after drinking alcohol.

Our reduced capacity to focus in these states means that our brains are full of other random associations and seemingly unconnected thoughts that we struggle to block out, and this is what leads to the insights. (Note: the research study on alcohol use shows that this affect only works with a small intake of alcohol!)

Similarly, there is plenty of anecdotal evidence (and even some research) of the use of other substances in creative work. Samuel Coleridge's exotic poem *Kubla Khan* was inspired by an opium-induced vision that he experienced.

Favourite Idea Generation Techniques

See Chapter 5 for a long list of idea generation techniques. Everyone has preferences amongst this list—experiment with the techniques and use the ones that work best for you. Note that many techniques aren't necessarily suited to every situation, and if a technique isn't working that well for a particular task, try another one rather than giving up altogether.

Location

Find your own space to do your creative work. Consider where you are when you do your most creative work or are at your most productive—and protect that space from invasion! The author Paul Auster uses a second apartment in Brooklyn, not far from his family home, to write. The separation from home is important, and his walk to the apartment gets him primed to write.

Research shows the powerful nourishing and restorative power of being in nature. How can you incorporate nature into your creative space? A window looking out onto the world outside? A pot-plant on your desk? A daily walk in a park?

Numerous artists and graphic designers pay testament to the power of being outdoors and surrounded by nature, not just for its positive effect on energy and demeanour, but also for the inspiration it provides in the form of shapes, colours, patterns and so on.

Graphic designer and artist Edwina Buckley said, "I love being around creation and seeing natural patterns that happen and the way things work together…the water and how the light reflects off that…and it can give you ideas for actually creating the artwork" (5).

Cafés have long been used as locations for creative work, particularly for those working in a freelance capacity, and it can be useful to consider whether a public location, with all its ambient noises—can help or hinder your creative endeavours.

A 2012 study published in the Journal of Consumer Research found that "a moderate versus low level of ambient noise enhances performance on creative tasks…a high level of noise, on the other hand, hurts creativity" (6).

There are now apps and websites designed to recreate the ambient noise of various locations, including coffee shops. So what effect does being surrounded by people without necessarily being interrupted have on our thinking? Do we work better with company, or in solitude? Is it important for us to get away from our homes or offices to engage in creative work? And what is the role of coffee in our thinking?

Environment

Consider the elements of the environment around you that might contribute to your creative capacity. Do you prefer to work in hot or cold temperatures, in light or dark rooms?

Certain companies are famous for the set-up of their offices, which employ games, toys, recreational equipment and a variety of stimuli. What external stimuli might be useful in your creative environment?

People You Are With

Are there people whose presence inspires you—their energy, or their encouragement, or their flow of ideas that spark your own?

Are there people with whom you collaborate particularly well? What are the features of these partnerships that work so well? What is it about working with those people that aids the creative process? There are numerous successful creative partnerships, not least in the song-writing world, where duos such as John Lennon and Paul McCartney, or Elton John and Bernie Taupin, have collaborated to write numerous songs that enjoyed huge success. Each brought complementary skills and strengths to the table, and the collaborative effect on creative output was striking.

How can you make sure you are regularly swapping ideas or working with certain people with whom you collaborate well?

The emergence and growth of 'incubators' and 'creative hubs', offices and organisations that bring entrepreneurs together in shared spaces, provide opportunities for people to work side by side, collaborate, and receive peer review and support.

Such incubators create an environment that is very conducive to creativity for some, and have shown considerable promise for quickening the pace of innovation. Because everyone is developing new ideas, there is a common bond, a recognition that however crazy or half-formed the idea, something good might emerge from it.

Diversity of Input

How can you solicit perspectives, opinions, and feedback from other people and thereby increase the diversity of input into your creative pursuits? How can you give yourself the opportunity to hear diverse opinions and harness other people's expertise? The watercooler and the lunchroom have often been the organisational meeting points where people converse and share opinions, reflections and ideas, and the design of office spaces can play an important role in encouraging creativity.

Can you pull together a team of people to work on an idea? What diversity do they bring to the table? As we'll see in Chapter 6, teams with diverse members are creatively more successful than homogeneous teams.

Build your networks and identify your supporters and collaborators. Seek multiple points of reference, collect different materials to consult and learn from, and look for new and diverse experiences that can broaden your perspectives.

Relaxation and Incubation

Make sure you build in periods of relaxation for incubation into your creative work (remember the process model of creativity in Chapter 1). Our understanding of the brain continues to develop, and we know that our subconscious will continue to work on a problem. Taking a break from thinking about it can help. However, don't use this as an excuse for prolonged procrastination!

Be mindful of which (regular/habitual) incubation periods yield positive responses, e.g., the morning shower, an afternoon run or walk in the park, a workout at the gym.

Painter Keith Bond (7) found that playing an instrument helped him to relax and, in due course, allowed the ideas that had been marinating subconsciously to come to the surface.

Nietzsche suggested that, "all truly great thoughts are conceived by walking" and research has shown that walking stimulates the brain. This would certainly justify all the walks taken by Tchaikovsky, Benjamin Britten and others that we described in Chapter 2.

According to Stanford University researchers (8), a brisk walk can increase our capacity to think creatively by up to 60 percent, as measured by tests to generate alternate uses for common items (e.g., the paper clip test). Certainly, exercise has a physical effect on the brain and leads to higher arousal and enhanced mood.

Creativity is increased when there is a strengthened connection and increased communication between both hemispheres of the brain. Many pre-schools and learning centres use music and *brain gym*[6] as ways to boost the connection and, thereby, creativity.

Props to Aid Creativity

What tools and props might help you in your creative pursuits? Would it help to carry a notebook or sketchbook for when an idea strikes? Or use the note pad or voice recording device on your smartphone to record ideas?

---

[6] *Brain Gym* is a programme of body movements used in schools to stimulate the brain for learning, though it has been criticised in recent years as being not scientifically proven to benefit the brain.

If you are trying to come up with creative ideas that would suit a particular target market (e.g., a marketing campaign for a target audience), what props can you use to put you in the mindset of that target audience? For example, you could listen to music that is popular amongst that audience or visit a particular bar or café that is similarly popular.

We've all heard of the proverbial 'thinking cap' to put us in the frame of mind to think effectively. A writer that we know has a 'thinking chair', his favourite battered old rocking chair. He sits in this chair to think and come up with creative ideas for his writing; for him, this chair is a prop to get him in the right frame of mind to think creatively.

Removal of Distractions

Find a location where you will not be disturbed. In many work scenarios, it can be hard to dedicate uninterrupted time to creative pursuits as offices are open plan, or people are frequently walking in and out of each other's offices. ("Sorry to interrupt but…"; "Can I just ask you a quick question?")

Remove other distractions from your location, particularly electronic ones. Turn off phone/email/social media notifications; resist the temptation and addictive urge to bury yourself in your social media, and take the opportunity instead to observe the world around you as a source of ideas as you commute on public transport, or while you are waiting for a friend at a café, bar or restaurant…etc.

Structure

Give yourself parameters, goals, or deadlines to help you get your creative work done. Contrary to the popular conception that structure constrains creativity, structure can be very helpful in encouraging certain aspects of creativity. As psychologist Rollo May commented, "Creativity arises out of the tension between spontaneity and limitations." (9)

Routine can help you persevere and continue to make progress with your creative efforts, while a lack of structure can actually be paralysing. The writer Elizabeth Gilbert talks about how she knows she just has to "turn up at the page every day" to make progress in her writing projects. The structure and routine that she puts in place to help her write is to simply sit down at the computer each and every single day.

Deadlines can be a source of motivation and focus, but those that are too frequent or tight, or indeed too few or lax, can also become demotivating and lead to burn out or lethargy and disinterest.

Goals and Targets

Using SMART goals can be really helpful in getting things done, and setting process goals can also be a very fruitful approach (see Chapter 4).

The Search for New Ideas

As we have discussed, the results of creative work are never entirely new, never the creation of something from nothing (10). These new ideas spring from the process of harnessing the ideas, products, processes, thoughts that are already out there in the world, and combining them, reformulating them, bending them into a new shape.

The tool company Black and Decker invited anyone to bring them inventions or ideas for new products. One day a man who was a keen gardener arrived at their offices with a curious contraption. It was an old manual hand-push lawn mower with the blades replaced by pieces of wire, and it had an electric drill bolted onto it to power it.

The man explained that it was for removing the moss and dried grass from a lawn so that new growth could come through. The enthusiastic young product managers took the contraption out to try it on the immaculate lawns that were lovingly tended by a retired Black and Decker employee. The 'lawnraker', as it was subsequently known, worked spectacularly well and took up plenty of moss...and unfortunately, a lot of lawn as well, much to the annoyance of the retired employee. (It was the job of one of the authors to smooth the ruffled feathers.)

Once the 'lawnraker' had been adjusted accordingly, it went into production and sold 80,000 units in its first year—providing a great example of the success that can arise from combining existing elements to create a 'new' product.

But how do we acquire the ingredients for these new ideas? We have already discussed the importance of knowledge and expertise in leading to creative ideas in particular domains. But so many of the issues, challenges and problems that we face, and that require creative solutions, pertain to our day-to-day lives and our existence as human beings building lives within our communities.

As a result, we can build vital knowledge and expertise to fuel our creative capabilities by observing the world around us and considering how we experience that world.

Observing the World Around Us

The world around us gives us the raw materials with which to be creative. It can provide us with new ideas and inspirations; materials and concepts to play with, use and combine in new ways; new connections; and observations of human truths that really resonate with other people and lead to creative ideas and solutions. These human truths show the true nature of human problems, and shed light on the path to effective and creative solutions. They also form the basis for artistic expression (e.g., Adele's emotional honesty).

Johannes Gutenberg famously applied his awareness of the screw presses used to press grapes and olives for wine and oils, together with his experience of metal stamping as a goldsmith, to invent the printing press in the 15th century.

The world around us provides so much material with which to work creatively, but unfortunately, we pay very little attention to it. Whether we're buried in our smartphones during a morning commute, or lost in thought worrying about a work deadline, we deprive ourselves of the opportunity to notice what is going on around us, and use those elements in our creative work.

Note that our *perception* of the world forms the reality of our experience, and our capacity to use our creative abilities. So, considering *how* our perceptions are formed can be impactful. Perception is made up of four elements:

- Exposure—what we are exposed to
- Attention—what we actually pay attention to
- Interpretation—how we make sense and meaning from what we pay attention to
- Memory—what we then learn from this experience, and how we use prior learning to influence our exposure, attention, and interpretation in the future

Once we have experienced something (exposure/attention), made sense of it (interpretation) and committed the experience to memory, we rarely undergo the interpretation process again. We just rely on our memory to provide us with the interpretation we made last time.

Whilst this is undeniably a very necessary process, as it makes our lives so much more efficient, this can be a major hindrance to our capacity to think differently and think creatively. Instead, in order to facilitate creative thinking, we must strive to cultivate an open mind and challenge assumptions and previous interpretations, as we will discuss below.

How we approach these four elements of the perceptual process can have an influence upon our capacity to be creative.

Exposure

Exposure is selective—whilst we are exposed randomly to a lot of stimuli (e.g., the people who are aboard the train with us on the morning commute to work), we also deliberately expose ourselves to a lot of stimuli (e.g., the TV shows we watch, the people we follow on social media, the websites we read). Conversely, there are also many stimuli we choose not to expose ourselves to (e.g., people or websites that espouse opinions contrary to our own, that we may avoid, or block or unfollow on social media).

By making deliberate efforts to expose ourselves to new people, places, information sources, stimuli—and by seeking new experiences (particularly those that intrigue us)—we can provide ourselves with a whole new set of materials with which to forge new combinations, that is to say, to be creative.

This, of course, can be challenging, particularly when we expose ourselves to important views that are strongly opposed to our own (e.g., moral, political). Nonetheless, these diverse perspectives are all useful fodder for our creative faculties.

Another challenge to broadening the scope of our exposure may be circumventing the algorithmic controls that govern so much of our online activity, from the posts we see on social media, to the suggestions we receive on YouTube or Amazon; these algorithms are designed to give us more of what we already consume—creating the bubble or echo chamber that has become a facet of online interaction—rather than serve us breadth and diversity of content. So it may take concerted effort to access material outside of your own 'bubble'.

Tactics that can help include reading outside your areas of interest, visiting new places, cultures, countries that are different to your own and talking to people who have a different background.

In 1979, the British band Joy Division released their album *Unknown Pleasures*, an album that would be heralded as a classic of the time. The front cover image, a graphical image in black and white, has since become iconic, replicated in all manner of ways on t-shirts, shoes, dresses and other items of clothing, household items, ashtrays, and even as tattoos.

But, as graphic designer Peter Saville explains, the image is actually a comparative rendering, taken from the *Cambridge Encyclopaedia of Astronomy*, of the frequency of a signal from a pulsar. (A pulsar is a particular type of star; apparently the first pulsar was observed by a research group at Cambridge University in 1967.) The band had come across the image and gave it to the graphic designer as inspiration (11).

How you use your vacation time can be impactful (12). A study of 46 Dutch workers found that after going on an international holiday for two to three weeks, they were able to generate more diverse ideas for alternative ways to use everyday objects, such as bricks, tires, spoons and pencils.

Researchers in Singapore have likewise found that greater exposure to other cultures through traveling, having international friendships, studying languages, and consuming music and food from other countries is linked to unconventional problem-solving.

Considering how to see your situation through another person's eyes can provide exposure to new experiences and subsequent insights. A charity for people with disabilities asked all its able-bodied staff to experience for a day what it was like to deal with a disability. Some took to wheelchairs, others were blindfolded, some went to immigrant communities to experience not being able to understand the dialogue around them.

This process exposed these staff to a new experience, opened their eyes to a different perspective, and shattered the assumptions that they might have held about the experience of being disabled. As a result, the staff were much better equipped to explore how to make their clients' lives better. It helps, therefore, to deliberately widen our knowledge of areas *that are outside* our normal experience. [7]

---

[7] A radio programme entitled 'Does he take sugar?' epitomised the wrong assumptions people make when faced with someone in a wheelchair. The title pointed to the common

Furthermore, research would suggest that doing new things, such as visiting somewhere new, or undertaking familiar processes in new ways, such as taking a new route to work, would actually lead to increased divergent thinking skills (when tested right after the new experience) (13).

Some techniques to bring about new exposure:

- Take a new route to work
- Visit a new part of town
- Take a vacation to a new destination
- Join a club to meet new people
- Watch a new TV show, read a new book, go to a website you've never visited before—particularly a different news website—follow someone new on social media
- Change the position of your desk
- Eat or drink at a different café, restaurant or bar

Attention

The step beyond exposure is attention. It's one thing to be exposed to something, it's another thing to actually pay attention to it. How much of what we are exposed to do we actually pay attention to?

Without looking…what is the app on your smartphone that is closest to the top right corner? Once you have decided on an answer, turn on your phone to take a look and verify. Did you manage to remember correctly? You look at your phone so many times every day—research shows that in the US people unlock their phones 58 times a day (14)—and yet it isn't necessarily easy to remember what this app is.

Now lock and put your phone down again. What is the time?

Most smartphones will have the time on the home screen or the lock screen, or both, so in the action of checking your phone to verify app placement, you will have been exposed to a clock. Yet most people are usually unable to recall the time.

Pay attention to what is going on around you. There's so much that we miss! Pay attention because what you see in the world around you is the raw material

---

assumption that everyone in a wheelchair is somehow incapable of thinking for themselves.

of everyday life, and a source of the knowledge to be creative within it. Our attention goes very much to where we focus it and ignores everything else. However, this can mean that we miss a lot of the detail of life around us that could be useful material for creativity.

When you walk down the street, look around you and take in your surroundings, rather than just looking at the pavement, or getting lost in trance-like thought (usually worry), or being glued to your smartphone.

A graphic designer was once talking with one of the authors, and was asked about his sources of inspiration. He paused, then pointed to the sunlight that was pouring through the blinds of the room and making patterns on the table. "See those shadows," he said. "The way they make shapes. The way they sparkle, and dance around. They give me ideas for my graphic design."

The author had not noticed or paid any attention to the sunlight, but the designer had—and was used to looking out for the elements occurring around him every day that might provide him with inspiration for his work.

An advertising executive, who had used a specific graphic effect in a commercial for a brand of bottled water, explained that the idea for the graphic effect had actually come from the opening credits of a movie he'd seen some time earlier. He noticed and paid attention to the graphic effect in the movie credits, realised that it was an idea that could be applied to some advertising at some point, and filed it away in his memory until an appropriate opportunity to use the idea arose.

And as we've discussed, creative output is not so much 'brand new' as a new combination of existing ideas, or a new perspective on, or application of, an existing idea. The first step towards this is seeing what other people don't. This will help you to think what other people are not thinking. Pay attention and be present to what's going on around you in the world, and you will acquire a heightened appreciation of the building blocks of everyday life with which to generate creative solutions to life's problems.

One morning in 2007, the acclaimed violinist Joshua Bell 'busked' in a Washington DC metro station for 45 minutes during rush hour. Most people ignored him, and didn't even look up, let alone stop and appreciate the music—of the 1,097 people that saw him that morning, only 7 stopped to listen. 27 people gave money, for a total of $32.17. Yet Bell's concerts regularly sell out, with ticket prices of over $100.

This famous experiment (15) illustrated the extent to which our attention can miss impactful things, and also the extent to which our attention, as well as the value we place on stimuli we are exposed to, can be guided by context and assumption. If we can challenge these assumptions and be more open to paying attention to what's actually going on around us, regardless of the context, we can profit creatively from what we encounter. Can we spot great ideas in unexpected contexts?

Tips for paying attention:

- Look out of the window while you are on the bus or train, rather than looking at your smartphone or reading a book, and you'll see what other people are not seeing, you'll notice things that most people won't.
- Take a social media break for a whole day!
- Meditate—this helps you improve at being present in the current moment, a necessary state in paying attention to what's going on around you.
- Choose an object to draw, and then draw it by attempting to convey every smallest detail about the object that you can possibly notice; this will allow you to practice noticing and really paying attention to the detail.
- Don't ignore what your senses are saying—the heart as well as the head. Be alive to what all your senses are telling you—sight, sound, smell, touch, taste and the sixth sense of intuition. What are you giving attention to? What are you ignoring?
- Be open and present to new experiences.

Interpretation

We invariably use existing paradigms of thought, often based on our past experiences, or upon prevailing societal wisdom ('the way it's done') to make sense of everything around us. This is a survival technique—to make sense of everything, and to do it quickly and efficiently—so that we can respond appropriately. Whilst this is an efficient way to conduct our lives, it can be an unproductive approach when we seek creative solutions.

We have a natural tendency to rush to judge and so make snap decisions. But it can be helpful to practice suspending judgement and resisting the urge to rely on familiar interpretations and instead to look for new ways to interpret what you see around you. Keep an open mind as long as possible.

There is a saying: "The shortest distance between two points is the long winding road." In other words, the road to breakthrough can be long and winding, so embrace the journey of discovery and don't be too anxious all the time to press ahead and miss out on important clues on the way in doing so.

There are two simple but powerful questions we can ask to help us do this:

- Why?
- What if?

Ask 'why?' over and over again (see the '5 whys' technique in Chapter 5) and the follow-up question, 'so what?', to help drill down to the deeper issues and human truths. We might encounter some new aspect of the problem we've never considered before, and this can inspire new connections and insights, and creative and effective solutions. Asking why is what children do, because they don't yet have a wealth of prior knowledge and experience with which to make sense of the world, and as a result they are going through the process of interpretation for the first time. [8]

Ask 'what if?' to challenge assumptions around the way things are currently done and open the door to a wide range of new possibilities that we don't usually consider.

At the matchstick company, Bryant and May, the story is told of someone on the assembly line who challenged the notion that there needed to be a striker strip on both sides of the matchbox. They simply asked the question 'why?' It turned out to be an assumption that there should be a strip on both sides, with no justifiable basis to the assumption. As a result, Bryant and May decided to place a striker strip on one side only, and greatly improved their profitability.

Challenging our assumptions, suspending judgement and resisting our habitual interpretations of events and experiences, can be vital to the creative process. Apple had an advertising campaign, entitled *Think Different*, which showcased a series of renowned 'creative geniuses' from Einstein and Picasso to

---

[8] In asking 'why?' of other people, one must be careful of the tone of voice used. Instead of conveying curiosity, the wrong tone can be interpreted as implying that the other person is not competent.

entrepreneur Richard Branson, aviator Amelia Earhart, and opera singer Maria Callas.

A voice-over recited a poem, beginning with the line, "Here's to the crazy ones" and praising "the misfits, the rebels, the trouble-makers, the round pegs in the square holes, the ones who see things differently". It makes the point that the world is changed, and the human race pushed forward, by those people who challenge assumptions, refuse to accept the status quo, and see things differently. It acknowledges that these people are often mistreated and called crazy. But they are the ones who are the creative geniuses who bring change to the world.

Consider art, music, film, TV, comedy, literature, life hacks; much creative output comes from observing the world—particularly human interaction and behaviour—and asking *why?* to truly understand what is going on. This resonates with all audiences.

Comedians often rely on this insightful observation of the human condition to create their material. British comedian Eddie Izzard, for example, found comedic material in asking 'what if?' questions, such as wondering how and where staff on the Death Star (in *Star Wars*) ate meals, and what if the Death Star had a canteen like many other workplaces. Or what if all animals were good and evil (like humans)—what would the behaviour of evil pilot fish or evil giraffes look like?

The British actor and comedian Ricky Gervais was asked about the biggest single influence on his creative process, in a series on Creation Stories for creativity website, Fast Company, and talked about noticing the everyday details.

He was told by an English teacher when he was 13 or 14: "Write about what you know." And implicit in that, Gervais explained, was noticing all the everyday details of that world you know, and including them in your creative output. This is the key to connection and to creative work that resonates with many other people—the ordinary human truths and the honest details of everyday lives (16).

Memory

We learn from experience. But creativity is all about the new, the unknown and the unexplored. So, we sometimes have to fight to ignore what our memories are telling us about how it was the last time we saw/did/said something, about how it *should* be. So:

- Stay curious—keep learning (see below), rather than just resorting to memory, or to the same old experiences
- Cultivate an open mind by deliberately resisting the urge to quickly judge
- Look at the world with a sense of wonder
- Consider why people act in a certain way, and why you respond in a certain way

Our memory is also the gateway to all the knowledge and expertise we have previously acquired that now serves as the raw material for new connections and new creations. So we must strike a balance between harnessing the memory of previous experiences that have brought knowledge, expertise and can be reconstituted in new approaches, with looking for ways to strengthen and add to our memory the new knowledge that will fuel future creative insights.

Remember also that memory is fallible. It can be subjective and distorting, and different people can have different memories of the same event. This may be because they have paid attention to different aspects of the event, or have interpreted it in different ways.

In some cases, particularly traumatic experiences, people may remember the details of an event very differently. Two roommates in Manhattan (one of whom is one of the authors) spent the day of 9/11 together. They were living together in an apartment near the World Trade Centre, and as a result were caught up in the thick of the day's events.

Later that day, and subsequently over the coming days and weeks, they compared notes on their experiences and found that they remembered certain details, such as the sequence of events, very differently, and even in contradictory ways.

So hold lightly to your trust in your memory of events, details and interpretations, and be prepared to explore, consider, and value other people's memories and the differing interpretations they hold.

A Simple Meditation for Being Present

- Close your eyes.
- Take deep breaths—in through your nose, down into your belly (feel your belly expanding), and out through your mouth.

- Feel yourself grounded on the chair you're sitting on. Feel your feet grounded on the floor. Feel your weight sinking down through you onto the chair.
- Hear all the noises around you: the traffic, the sounds of nature, the sounds of other people going about their daily lives.
- Feel your shoulders drop.
- Keep breathing deeply.
- Imagine your head being pulled up from the top by an imaginary force.

Learning

Learning is crucial to building the knowledge and expertise that is an ingredient of creativity. Consider how you go about acquiring this knowledge. What is your approach to learning? To what extent do you seek out opportunities to learn? What are you learning today? What are you *not* learning? How are you continuing to hone your craft as well as explore novel areas?

The considerations around observation, exposure and attention that we've just discussed are crucial parts of the learning process. Many of the idea generation techniques we discuss in Chapter 5 will also enable learning. However, there are some other elements that are worth considering.

Learning Styles

When we consider the conditions for learning we know from studies of accelerated learning that stimulating all the senses helps.

A number of psychologists have conducted research into the way we learn, notably David Kolb whose work on learning styles was developed by Peter Honey and Alan Mumford (17) into a process to identify which of four learning styles we thrive under. We all have an individual preference for learning in one of the following four styles:

- *Active* learning—learning through active experimentation and trial and error.
- *Theorist* learning—learning through understanding the underlying theory before experimentation.

- *Reflective* learning—learning through watching and reflecting upon successful role models or examples before practicing.
- *Pragmatic* learning—learning through understanding the practical application of the learning, and putting it immediately into action.

When we are able to use our dominant style (or whatever combination suits us at the time) we are more likely to stay motivated, remain curious, continue exploring and questioning in order to understand, and more effective learning takes place. This learning then becomes raw material for our creative endeavours, accessed via our memory.

Learning from Others – Benchmarking

We can learn a lot from other people. Be curious about other people's experiences and consider what you can learn from them yourself. The knowledge that you can glean from someone else's experience may be the information that provides fresh insight and innovative solutions.

Look what others do, in other industries, fields, cultures, countries. This provides exposure and access to new information on how to do things.

Trawl through old ideas that could not be used at the time due to old technologies or old mindsets. Patents from Victorian times often displayed great originality, but the technology of the time couldn't turn them into practical solutions. Today's technology might be able to.

On an organisational level, it can be very useful to look at what competitors are doing, and consider what they are doing differently. This is a process of benchmarking, and it is sometimes discounted as an approach because we have a limited view of the companies from whom we might learn.

Thinking beyond direct competitors to other more tenuously related companies and industries can be useful. An airline wanted to examine its telephone reservation service and sought to benchmark the process. Rather than going to a competitor company; i.e., another airline, which would be difficult given the competition between them, they went to an entertainment arena instead and picked up valuable ideas about staffing profiles at different times of the day/week/month.

Competitors in any industry tend to copy each other, so looking to organisations outside the industry can be a source of new perspectives and approaches.

Learning Through Dialogue

Language and discourse shape our thinking and vice versa, so what is the nature of discourse that facilitates creativity? Whilst it is certainly true that many new ideas come in a solitary moment—Archimedes in his bath and Newton under his tree—many also come from the synergy that arises from interaction between people. There are three components of discourse to consider:

a. how we listen
b. what we say
c. how we encourage a creative conversation

How We Listen

Encouraging listening	Discouraging listening
Curious and open mind	Only receptive to what is already known
Seeking to understand the other • Insight questioning • Exploring ideas and assumptions • Summarising to gain clarity	From my perspective—agree or disagree • Stating my position • Defending my position
Integrating their view and mine	Determining whether right or wrong
Looking for a new approach	Looking for the flaws
Practical application	Assuming 'it won't work'
Listening to the words and the music • Choice of words • Tone and inflection • What is not said	Listening only to the words
Time to digest and consider, non-judgemental	Rush to judge

The basis of encouraging listening might be summarised by the old phrase: 'If you wish to be understood, seek first to understand.'

What We Say

Plain speaking is a useful ground rule in any forum, but a caveat is that speakers should consider not just what they want to say but how to say it so it can be heard and taken on board. Different cognitive preferences (see Chapter 8) might demand tailoring your message to different styles. For example, will data and logic be more effective than pictures, music, stories and anecdotes, or less so?

Encouraging Language	Discouraging Language
'Anybody got any ideas?'	'We tried it before'
'Before we decide let's look at all the options again'	'It would take too much time'
'I am keeping an open mind/I have changed my mind'	'I haven't got time right now'
'Let's try it and see if it works'	'It would cost too much'
'Nothing ventured nothing gained'	'That's not my job'
'I don't understand why we cannot we explore it further'	'That's not your job'
'I know we have different ideas, but let's see if we can put them together'	'That's not how we do it here'
'How could we improve? What else could we do?'	'Why don't you put that in writing'
'What would happen if...?'	'It's impossible'
'What have we missed?'	'You may be right, but...'
'Let's just imagine for a moment'	'That's stupid'
'Why do we always do it that way?'	'Our customers / managers / staff would never agree to that'
'Why are the rules like they are?'	'You cannot do that here—it's against the rules'
'Let's challenge the current method'	'I don't think that is important / relevant / helpful'
'OK so we got it wrong, let's see what we can learn'	'Ideas from those people don't count'
	'It's good enough'
	'If it's not broke don't fix it'
	'Our company is too small/too big'
	'Any mistakes and you will be punished'

Language often uses various devices to explain concepts. We use simile ('He went through that team like a knife through hot butter'), metaphor ('She is fishing in dangerous waters') and a combination of words to give a more graphic understanding ('well-oiled machine'—indicating efficiency and smoothness of operation'; 'basket of fruit'—a team whose members are a collection of different talents at various stages of ripeness).

By varying the combinations, we generate different mental pictures that might stimulate different thinking (see Chapter 5 for more examples of metaphors).

How We Encourage a Creative Conversation

A basic survival mechanism lies in our ability to make quick decisions—fight or flight in times of danger. However, this can often translate into a rush to judge, and with our negativity bias (tendency to take a negative perspective over a positive one), lead to the rejection of creative ideas (18).

Apparently, Steve Jobs' first response to the script for the 'Think Different' commercial we described earlier was one word: "shit". In addition, something very positive will generally have less of an impact on a person's behaviour and cognition than something of equal emotional substance that is negative, which also increases the chance of rejecting creative ideas and opportunities.

A further negative influence lies in the confusion that can arise when people mix statements about: what is, what should be, what could be, and—if there were no constraints—what might be. Clarifying the intended meaning can help keep the conversation in the right mode (see 'modes of thinking' in Chapter 5). Talking at cross purposes across these different modes of speech can cause frustration and inhibition.

Successful creative conversations go with the flow of an idea rather than introducing a critical evaluation too early. Some years ago, a power utility company was faced with the problem of wrecked power lines brought down by heavy snow in locations that were inaccessible for repair crews. As the company prepared for the next winter, a group discussed how they could deal with the problem when it arose.

Someone suggested: "Why don't we get the bears in the woods to knock the snow off?" A natural response might have been 'what a stupid idea', or 'that's impossible'.

Instead someone else said, "Great, we could get them to climb up the pylons."

Another said: "Yes, we could put honey on the pylons and that will attract the bears."

"Super idea. Let's use helicopters to pour honey over the pylons."

"Hold on; if we fly a helicopter above the pylons and lines, it will blow the snow off, so we don't need the bears or the honey."

The members of the group used encouraging language, suspended judgement and built on each other's fanciful ideas until a practical solution emerged.

Behaviours and dialogues that help the encouragement of idea development include:

From this behaviour…	To this behaviour
Quick to judge	Positive encouragement for silly ideas
Stating only my view	Listening to words said and unsaid. Summarising and reflecting
Using only previously tried methods and solutions	Being prepared to experiment Working with colleagues with different backgrounds and experience Searching widely for solutions in different fields
Accepting/Rejecting evaluations	Keeping options open—'por que no los dos' (Why not both)
Strictly rational	Emotional and rational

Finally, creativity, almost by definition can stray into the wild and wacky, and one approach to encouraging this aspect of creativity is to disregard any consideration of 'utility' in the resulting ideas.

One example of deliberately freeing oneself from utility is 'Chindogu' (19), a Japanese term for inventions that are man-made objects that have "broken free from the chains of usefulness. They represent freedom of thought and action: the freedom to challenge the suffocating historical dominance of conservative utility; the freedom to be (almost) useless". The aim is initially to free thinking of the constraint of utility, introducing such considerations later on.

Chapter Summary and Conclusions

In this chapter, we examined several ways to enhance our individual capacity for creativity.

- Factors that can impact us individually include: time of day, energy cycles, idea generation technique, location, environment, other people, diversity of input, relaxation, props, removal of distractions and structure.
- By increasing our awareness of those factors, we can replicate the conditions that will help us be at our most creative.
- By observing the world around us we can find new ideas, for example, by benchmarking with other organisations.
- Determining how we learn and how we can effectively engage in creative conversations all help to create conditions for breakthroughs to occur.

Chapter 4
Navigating and Overcoming the Blocks and Barriers to Creativity

Introduction

Creativity is the life-blood of the creative arts—and the heartbeat of business in the 21st century. The successful IPOs of organisations like Facebook and Snapchat, or the sale price of Instagram and WhatsApp, demonstrate just how much is at stake from an innovative idea. The business literature is rife with discussion on enhancing the creativity of the workforce.

As we saw in Chapters 2 and 3, creative output relies on the presence of three key elements: knowledge/expertise, creative thinking skills that harness imagination, and an appropriate attitude that includes intrinsic motivation (1).

Accordingly, many approaches to improving creativity focus on these three ingredients of the creative process in the following ways:

- developing and increasing knowledge and expertise in a particular subject matter or domain through education, research, curiosity, trial and error.
- enhancing creative thinking skills and use of imagination through the practice of idea generation techniques (see Chapter 5).
- and finally developing a mindset that includes a positive open attitude and intrinsic motivation.

There is, however, an element of our attitude that is integral to our creative success, and that is often overlooked: our capacity to persevere. The assertion by Thomas Edison that "Genius is 1% inspiration and 99% perspiration" is a well-worn cliché, but with good reason.

For genius—in the form of creative ideas—to have value, to come to life, to make a difference in the world, we need to see these ideas through to their completion, and there is plenty of arduous work involved. This requires significant perseverance, as well as the intrinsic motivation and appropriate attitude to fuel and facilitate this perseverance.

Having a clever idea may be the easy bit (relatively speaking); it's the follow through that poses the challenge. The average novel contains about 80,000-100,000 words of carefully crafted, reworked and edited prose, a process that requires significant determination and perseverance. And it's this very process of generating the 80,000 words to bring an idea to life that prevents many great ideas from coming to fruition in a completed novel.

In 1991, Metallica became the biggest metal band in the world with the release of their self-titled 5th album, aka 'The Black Album'. In the aftermath of its monstrous success, producer Bob Rock commented (2): "The Black Album was the result of hundreds of hours of recording; the entire process lasted well over a year. It took them three months of work just to record the guitars. Editing the drum tracks took six months and legends abound regarding Lars taking weeks to find just the right snare-drum sound." The album was a masterpiece for the band, but it took some serious dedication and perseverance to see their ideas come to fruition in the way they wanted, and the way that would lead to such success.

Unfortunately, there are a number of all-too-familiar factors that undermine our capacity to persevere and pursue an idea to a practical outcome. More often than not, we have the best of intentions, but these factors derail us and send our creative projects grinding to a halt.

The synopsis for a novel that is never turned into a full draft, the idea for the promotional stunt that never progresses past the meeting room, the clever invention that remains a napkin sketch, are the all-too-familiar consequences.

The novelist Graham Greene had already enjoyed numerous successes as an author when he found himself dealing with writer's block—a creative "blockage" as he called it. He found himself unable to develop, or even start, stories, and as a result, was stuck and unable to write any books.

However, the process of keeping a dream journal, something that he'd first undertaken during a bout of psychotherapy as a sixteen-year-old, was his saviour, providing freedom from his "conscious preoccupations" (as he termed them in speaking to a friend) and freeing up his capacity to write again (3).

Julia Cameron was a successful journalist and screenwriter, when she hit her own creative wall, plagued by alcoholism and drug addiction. She found that she needed to develop an approach to unlocking her creativity again. As she explored pathways, she discovered the route that led to her bestselling guide, *The Artist's Way* (4).

For most of us, the wall we hit is not as severe. Instead, it's a bout of procrastination, or a lurking fear of failure, that trips us up and gets in the way of our creative endeavours. But these hurdles all add up, and since many creative endeavours can require us to keep working on a project over time, they can ultimately defeat our creative efforts. Unfortunately, in the realm of creative practice, successful results are often a case of 'easier said than done'.

Accordingly, we must develop strategies that combat these issues and help us persevere. To do so requires recognising the various ways that we can be challenged, and then employing various techniques to overcome them so that we can persevere with our creative work.

This chapter will examine the nature of these factors, and discuss approaches to dealing with them.

The Blocks and Barriers to Achieving Our Creative Goals

There are many challenges in maintaining the right attitude and mindset in order to persevere with creative practice so that we can achieve our creative goals and see our ideas come to fruition.

Among the most familiar are the following:

- Fear of failure
- Perfectionism
- Procrastination (which is how fear of failure and perfectionism play out)
- Being too busy
- Being too tired
- Being too bored
- Feeling uninspired and being passive
- Doubting our own capacity to create anything of value (self-doubt)
- Fear of being 'found out' (imposter syndrome)

- Criticising ourselves too harshly
- Fear of early success (as it might imply future commitment and hard work)
- Negative self-talk
- Experiencing failure, criticism, or setbacks early in the process

As a result, we create self-limiting beliefs about our capabilities and then use these to prevent ourselves from being creative.

George R.R. Martin, author of the *Game of Thrones* series of books, commented that, "It's like constipation. And you write a sentence and you hate the sentence, and you check your email and you wonder if you had any talent after all? And maybe you should have been a plumber?" (5)

Procrastination provides the excuse we need to avoid that fear of failure or the challenges of perfectionism. And when it paralyses all efforts to undertake creative work, we must consider how to address it. Even when the work gets done, procrastination can lead to some very uncomfortable and pressurised experiences in completing the work, as illustrated by the humorous graphic below:

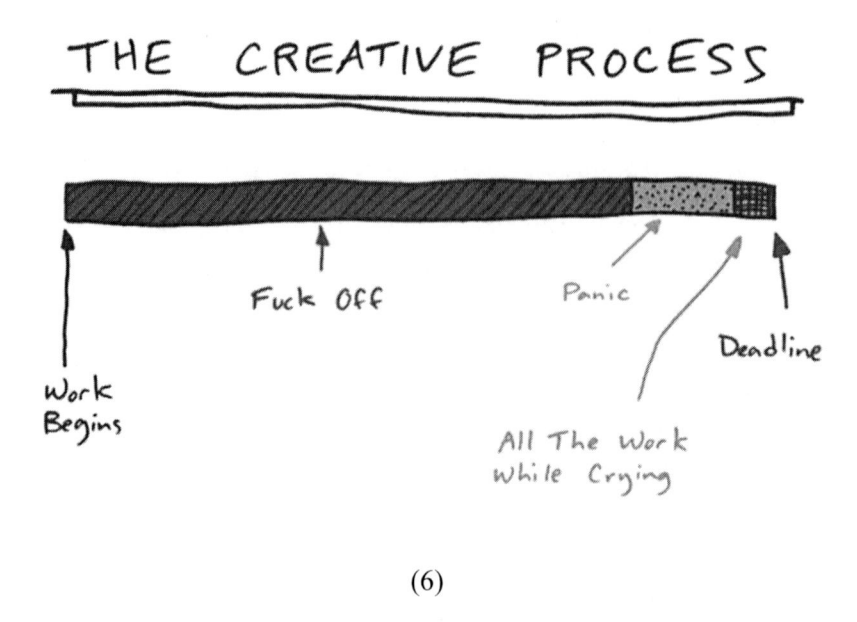

THE CREATIVE PROCESS

Work Begins — Fuck Off — Panic — All The Work While Crying — Deadline

(6)

Sometimes, the impending deadline is the only thing that will galvanise us into action, but our procrastination might not be a wilful neglect of the creative practice, so much as an attempt to deal with other more pressing and timely issues in life.

Furthermore, procrastination may actually also include time for day dream and for incubation, and as such, may be beneficial to the creative process.

The extent to which we tolerate procrastination, or willingly allow it to have a place in our creative process, will depend on how we work best. However, if it becomes clear that procrastination is making creative work very challenging or impossible to finish, then it might be worth examining the underlying causes (e.g., fear of failure) and considering how to address it.

In his book *The War of Art*, writer Steven Pressfield (7) gathers together all these mental hurdles to our creative undertakings under the umbrella term of "Resistance": resistance to putting in the work to complete a creative project, resistance to taking the risk of putting that project out into the world, resistance to paying heed to one's creative urges and one's creative dreams…

So, what are the techniques to beat resistance?

Grit

University of Pennsylvania professor Angela Lee Duckworth (8) has pioneered the research into the psychological construct of grit. By her definition, grit equals perseverance and passion for long term goals. She conducted a series of research studies and her findings suggested that grit is actually the best predictor of success in life—a better predictor than IQ, EQ, social background, education and so on.

In her research with a group of US Army cadets, for example, she found that scores of grit were better predictors than anything else of whether the cadets would complete a rigorous summer training programme.

So, grit is a measure of the extent to which we can persevere to the successful achievement of long-term goals that we feel passionate about…such as creative projects. The more grit we have, the more likely we are to succeed in those projects[9].

---

[9] There is some debate about whether grit is the same as conscientiousness (a personality trait whose level remains fairly stable), but Duckworth maintains that it is within the power of all of us to increase our levels of grit.

WD40, a penetrating lubricant that is widely used to free up bolts and mechanical devices of all kinds, was originally developed in the missile industry for water displacement purposes (hence WD). Iver Norman Lawson allegedly took 40 attempts to develop it, and demonstrated his grit and determination in the process!

Some techniques to increase grit that emerge from research by Duckworth and others include:

- Pursuing what interests you. Those with a passion, as well as a clear vision of what they could become once they achieve their dreams, tend to be more resilient, especially in times of failure and rejection.
- Finding a purpose in what you are doing and determining how it is meaningful to you and your life. In his book *Drive* (9), Daniel Pink identifies purpose (along with autonomy and mastery) as a key driver of motivation in any sphere of life. Reflect on your intrinsic motivation. Why are you doing this task/project? What are the underlying reasons that make the task/project worthwhile for you?
- Holding on to the hope that you can make a positive difference in the world with your work. Those with hope tend to explore and experiment to find a way forward whereas those with little hope of success tend to confine themselves to task-oriented goals or give up.
- Using meditative practice. Running is a powerful meditative practice for one of the authors, something that centres him.
- Engaging in spiritual practice. This is a powerful tool for the other author.
- Cultivating support groups—family, friends, colleagues.
- Taking care of your physical fitness and wellbeing.
- Having good role models.
- Having a moral compass.
- Using positive language—to ourselves and others!

Mental Toughness and Resilience

Resilience is defined as "the capacity to recover quickly from difficulties", which is exactly what is required the first time the product of your creative endeavours is rejected. Of course, the creative arts are littered with stories of creatives who bounced back from multiple failures. JK Rowling's manuscript for *Harry Potter* was rejected by 12 publishers before Bloomsbury picked it up.

Yankee Hotel Foxtrot, alternative rock band Wilco's most acclaimed and successful album, was rejected by the band's label Reprise Records. Wilco reacquired the rights to the music, left the label and released the album through Nonesuch (ironically owned by the same parent-company—Warner—as Reprise) instead.

Both became success stories—but they could have easily given up or abandoned projects after rejection without the requisite self-belief, resilience and mental toughness to persevere. Margaret Thatcher famously remarked that she believed that there would never be a British woman prime minister in her lifetime. But her political perseverance and mental toughness ultimately prevailed and she became that first woman prime minister in the UK.

During the second world war, a British inventor, Barnes Wallace, invented a 'bouncing' bomb that led to the breaching of a German dam. Despite numerous failures in his attempts to produce this bouncing bomb, and scepticism amongst the military leaders as a result, Wallace persevered because he believed that destroying that dam would cripple military production and so shorten the war. It was his resilience, borne out of strong belief that the goal was worth the effort, that gave him the impetus to persevere until his bouncing bomb was successful.

Resilience comprises a collection of different abilities and faculties. Coach and author Carole Pemberton (10) identified eight components of resilience, some of which dovetail with grit, and all of which can be developed. They include:

- Confidence in your self-efficacy (your capacity to get things done)
- Flexibility to adapt to changing circumstances
- A sense of purpose that all the effort is worthwhile
- Creativity in looking for different solutions
- Support from others
- Proactivity to face the issue

- Emotional control to retain perspective
- Realistic optimism (see the Stockdale paradox[10])

Some Techniques to Enhance Resilience

There are several ways to build greater resilience. Some of these may seem obvious or not necessarily applicable, but the key is to identify approaches that are likely to work well for you as part of your own personal strategy for fostering resilience. Below are some tactics that can help in the quest to develop the components of resilience listed above.

Be clear about your reasons for working on this project and the benefit that might arise from successfully completing the project. Don't be put off by early failure or be complacent about early success. There may well be criticism early in a project, but it may be completely misplaced.

If you are clear about why you are working on a project, and what the positive benefits can be, it becomes easier to ignore this criticism and push ahead. On August 17, 1902, *The Post* newspaper called bicycling a passing fancy, and declared that "the popularity of the wheel is doomed". Critics thought bikes were unsafe, impossible to improve, and ultimately impractical for everyday use. Nonetheless, the bicycle prevailed!

Make connections with others, and build and nurture a strong support network. Good relationships with close family members, friends, colleagues or others are important. Make sure you know who you can turn to for support, and make sure you nourish those relationships. Accepting help and support from those who care about you and will listen to you strengthens resilience.

Some people find that being active in civic groups, faith-based organisations or other local groups provides social support and can help with reclaiming hope for a positive outcome. Assisting others in their time of need can also benefit the helper. In a study of self-employment by one of the authors, the positive role of a circle of significant others was a critical factor in determining success.

---

[10] The Stockdale paradox—a balance of optimism with realism—arose in the Vietnam war, taking its name from Admiral Stockdale who was held in the notorious 'Hanoi Hilton' prison as a prisoner-of-war. Other prisoners believed they would be rescued from captivity very quickly and when this didn't happen, they lost hope and gave up trying to survive. Stockdale believed he would survive but would not be rescued for some time. He survived because his optimism was balanced with realistic expectations.

Avoid seeing crises as insurmountable problems. Think of previous times you've successfully handled adversity, and recognise that you have the capacity to handle crises and that you will survive. You can't change the fact that highly stressful events happen—for example, when your creative idea gets rejected—but you can change how you interpret and respond to these events.

Try looking beyond the present moment to how future circumstances may be a little better. Note any subtle ways in which you might start to feel better as you deal with difficult situations. Use positive language to describe what you can do, as opposed to negative language around what you cannot do.

Accept that change is a part of life. Certain goals may no longer be attainable as a result of adverse situations, and you may need to change your aspirations or expectations accordingly. Accepting circumstances that cannot be changed can help you focus on circumstances that you *can* alter. Learn to live with the problem and navigate it appropriately.

Move towards your goals. Develop some realistic goals. Do something regularly—even if it seems like a small accomplishment—that enables you to move towards your goals. Instead of focussing on tasks that seem unachievable, ask yourself, "What's one thing I know I can accomplish today that helps me move in the direction I want to go?" Some familiar sayings sum up these ideas: 'eat the elephant in bite-size chunks,' 'a journey of a thousand miles starts with one step.'

Take decisive actions. Act upon challenges and adverse situations as much as you can. Take decisive actions, rather than detaching completely from problems and stresses and wishing they would just go away.

Look for opportunities for self-discovery. People often learn something from the experience of failure, particularly something about themselves, and they may find that they have grown in some respect as a result of their struggles. Many people who have experienced tragedies and hardship have reported better relationships, a greater sense of strength even while feeling vulnerable, an increased sense of self-worth, a more developed spirituality, and a heightened appreciation for life.

One of the authors worked with people who had been laid off from their jobs, and found that once they had found a new job, many of them reported that they had learnt a lot about themselves and felt stronger and more confident because of the experience.

Nurture a positive view of yourself. Developing confidence in your ability to solve problems and achieve goals helps build resilience. Building positive belief in your abilities and instincts from considering past successes, whether in times of adversity or not, reinforces a positive view of yourself.

A success log may be useful here to remind you of your past successes so that you don't forget them. As you reinforce the mental link between any current challenges you face and past evidence of when you have overcome prior challenges, this capacity for self-belief is strengthened.

Keep things in perspective. Even when faced with very painful events, try to consider the stressful situation in a broader context and keep a long-term perspective. Avoid blowing the event out of proportion ("it's not the end of the world"). Practice gratitude by regularly naming the things for which you are thankful, even if they are simply small daily occurrences. Cultivating a grateful outlook can place challenges in perspective, help to manage stress, and build resilience.

Maintain a hopeful (but realistic) outlook. An optimistic outlook enables you to expect that good things will happen in your life. Try visualising what you want, rather than worrying about what you fear.

Take care of yourself. Pay attention to your own needs and feelings. Engage in activities that you enjoy and find relaxing. Exercise regularly. Taking care of yourself—being 'properly selfish'—helps to keep your mind and body primed to deal with situations that require resilience. Remember the mnemonic: SHED—Sleep, Hygiene and Health, Exercise, Diet.

Find the silver lining in the experience. Focus on any unintended benefits that might arise from a particular situation. When John Kellogg and his brother Will accidentally left a pot of boiled grain on the stove for several days, the resulting mess became the launching point for their famous breakfast cereal, Corn Flakes.

Find the learning in the experience. As Henry Ford said, "Failure is simply the opportunity to begin again, this time more intelligently." No experience is a total waste, there is always something to be learned. Use this concept as a rebuttal to negative self-talk by replacing thoughts of 'I'm a failure' with 'I have learnt a useful lesson and will now try something different'.

Experiment with other behaviours and techniques to strengthen resilience. For example, some people find that writing about their deepest thoughts and feelings related to trauma or other stressful events in their life can help. For others, meditation and spiritual practices can help build connections and restore hope.

Use CBT Approaches to Address the Fear of Failure and Negative Self-Talk

Cognitive Behavioural Therapy (CBT) is a psychological technique designed to challenge faulty thinking (and resulting behaviours) brought about by an activating event. The catastrophising and negative self-talk that many of us engage in when we think about putting our creative endeavours out into the world are often the result of faulty thinking.

As an example, the activating event might be working on a creative project or submitting a song or piece of writing. The faulty belief is that it will be rejected and that everyone will think the author is a disastrous creative. And the consequential emotion and behaviour is fear and panic, and a failure to complete the creative task.

Using CBT, we could dispute the belief by asking questions like: "Will that really happen? Is this a rational belief? If it does get rejected, is everything else I ever do really doomed? If it gets rejected does that really mean I'm a worthless writer?" The answers to these questions will start to uncover the extent to which our fears are not realistic, and are simply the product of faulty thinking.

Finally, we can exchange the thought for another—"I cannot control how other people receive my art, but their judgement does not diminish my capacity as an artist, nor does it reflect on my worth as a human. In fact, it may well have nothing to do with me."

Recently, this approach has been popularised by the 'Act as if' movement (11). A typical approach focusses on:

- Changing the self-talk through generating particular (positive) thoughts, as if they are our natural, spontaneous thoughts.
- Changing energy patterns through generating particular energy tones, as if they are our natural, spontaneous emotions and feelings.
- Turning mind and imagination to generate particular images, as if they are the natural, spontaneous images of our imagination.
- And then taking action, as if we are competent and successful at the task.

Reframe a Fear of Failure

Fear is a very powerful force, particularly when it pertains to our own physical and psychological safety. A natural human trait is to crave safety and abhor risk. Going against the status quo and suggesting something new can be terrifying as a result. We have an internal mechanism—the censor—that prevents us from saying or doing anything that might carry a risk of failure or of making us look foolish or inviting social rejection.

This means that we shy away from experimenting with the new untested ideas that creativity encompasses. Furthermore, the uncertainty inherent in creative output on account of its novelty—there's no way of knowing how a creative idea will be received because it has never been expressed before—heightens the fear of failure and, in combination with perfectionism, which seeks after the 'right' and 'perfect' solution, leads to excessive tinkering with an idea or, more likely, complete inaction.

A fear of failure, and the accompanying perfectionism designed to keep us safe, often play out in procrastination, whereby we allow ourselves to be distracted from our creative pursuits by other tasks, and in doing so, avoid the confronting and terrifying prospect of completing work that might be less than perfect and of having to risk putting it out in the world.

We often justify this path to ourselves—"Oh, I'm not creative; I don't have time to spend generating ideas; the washing really needs to be done first, etc."—and we use these self-limiting beliefs to prevent ourselves from being creative.

A fear of failure may always be present in some capacity, but instead of shying away from this fear and seeking to avoid it, we might embrace its presence by interpreting it in a different way: as simply providing a marker that we are undertaking important creative work.

As psychologist Susan Jeffers wrote (12), "Feel the fear and do it anyway." Perhaps we can learn to see failure as a necessary part of the experimentation process, as a necessary step along the way. Failure helps point the way towards success (hence the advice to 'fail fast'), and scaling the heights of creative success cannot happen without failure along the way.

Ask the question—what can I learn from this failure? Many writings on efficiency and productivity suggest tackling the task we are most afraid of, or prone to procrastinate upon, first before any other. Self-help author Brian Tracy labelled those challenging tasks as 'frogs', and encouraged his readers to 'eat their frogs' first thing in the morning, referencing the apocryphal Mark Twain quote, "Eat a live frog first thing in the morning and nothing worse will happen to you the rest of the day." Take that challenging task and get it out of the way first! [11]

In a coaching session, a client who feared rejection and was unable to be an effective salesman, was set the goal to collect several rejections from prospective clients before his next session. He went to see a number of sales prospects but failed to get any rejections; instead he made several sales! He came to see that rejection was less common than he feared and that when it did happen, it was outweighed by the sales he acquired.

Basketball star Michael Jordan famously said, "I've missed more than 9,000 shots in my career. I've lost almost 300 games. Twenty-six times, I've been trusted to take the game-winning shot and missed. I've failed over and over and over again in my life. And that is why I succeed." He confronted any fear of failure head-on and took the risk that he might fail. And he did fail, many times. However, he also succeeded many times too, and in doing so, became a legend of the game.

Nonetheless, we also acknowledge that sometimes the fear can be well founded, and that there can be repercussions to proposing the radical ideas that creative perspectives are sometimes viewed to be. As a result, pursuing creative endeavours can sometimes encompass great bravery.

Value and Praise the Effort, Not the Outcome

Research shows that students who are praised for the *efforts* they make in solving puzzles, rather than for the amount they get right (13), then go on to attempt to solve more puzzles and harder puzzles than those who are praised for the number of correct answers. They take risks.

As collaborators in creative work, we must remember to praise the efforts others put into their creative endeavours, rather than only praising the outcomes (and any success that outcomes achieve), and in doing so, encourage them to take

---

[11] Whether Mark Twain actually said or wrote this has been disputed, but the sentiment is certainly pertinent.

creative risks. And when we consider our own efforts, we must remember to value the efforts we put into our practice, rather than simply fixating on the outcome.

Placing value on effort rather than outcome can also be a powerful tool against procrastination. Rather than placing pressure on ourselves to reach a particular outcome within a time-frame (e.g., I must finish this chapter of writing in the next 2 hours)—with the result that we then respond to the pressure by procrastinating and avoiding the task—we could place emphasis on the process instead without any pressure on the outcome (e.g., I will just write whatever I can in the next 2 hours).

Goal-Setting: Smart Goals, Process Goals and Sundowners

There are several approaches to setting goals that can be very useful in tackling procrastination and making progress on creative projects.

Considerable research demonstrates the effectiveness of SMART goals in getting stuff done and in thriving in life more generally (14). Using SMART goals in the pursuit of creative achievement can be powerful and can help prevent individuals from getting derailed in the slog to see the creative process through.

SMART goals are:

- Specific and Stretching—vague goals lead to vague attempts to achieve them; difficult, though achievable, goals lead to higher achievement
- Measurable—you need to be able to evaluate progress towards success
- Attractive and Attainable—if you don't want it, or can't achieve it, you're unlikely to put in sustained effort
- Realistic and Relevant—achieving the goal will bring some beneficial outcome
- Time-framed—you must have an appropriate time-frame in mind to motivate your efforts

Professors Edwin Locke and Gary Latham have examined the power of goal-setting, and quote research which demonstrates that setting and working towards valued goals is related to a greater achievement of outcomes, enhanced psychological well-being (15) and even better health (16)!

This is increasingly the case when you have set the goals yourself and are intrinsically motivated to achieve them, when they are meaningful, authentic and aligned with your values, and when they are challenging, but also specific and measurable (17).

In addition, using process goals rather than output goals can enhance perseverance. As the names suggest, process goals focus on the *process* of doing the work, rather than the *output* of that work. (This aligns with the suggestion above to value the effort (i.e., the process) rather than the outcome.) So, a process goal might be to sit down with one's guitar for an hour and work on song writing.

An output goal would be to write a song…which creates greater pressure around the quality of the output—is the song finished? Is the song good enough? and so on—that often leads to procrastination. A process goal just requires sitting down with the guitar for an hour, regardless of output, which feels a lot easier to achieve.

A variety of writers, such as Steven Pressfield, Julia Cameron, and Elizabeth Gilbert, use this technique and champion it as a way of beating that fear of the blank page. When Stephen King talks of writing in the morning and editing in the afternoon, the implication is that there is no pressure to be *good* in the morning, just simply to *write*.

The period of judgement comes later on—in the afternoon. Beat resistance (and defer the pressure of success) by just sitting down at the page. That's how the work gets done.

Process goals are also much easier to incorporate into routines—30 minutes of writing every morning for example. The process—the journey—can be the key to unlocking the creative outcome.

Sundowners are a list of tasks—and there should be only two or three tasks at most on this list—that you will achieve before the end of the day, *no matter what*. For this to be effective, each task needs to be achievable—not something that will necessarily take you hours, or requires input from someone else. By getting at least a few things done each and every day, no matter how simple those tasks are, you maintain momentum, and slowly but surely make progress towards completing your creative project.

Other Techniques

When we start thinking about approaches and hurdles to creativity in this way, there are also many techniques and practices from the field of psychology that could be harnessed in the creative process. Some of the reasons we hold a fear of failure, a fear of looking foolish, a fear of going against the grain, may have their origins in past wounds on a psychological level (e.g., childhood experiences) that could be helped with psychological techniques.

This may entail working with a trained psychologist, therapist or coach, or using techniques that can be employed on your own. For example, energy psychology practices like EFT (emotional freedom techniques) have been successfully used with trauma amongst war veterans (18), and could be used individually to treat much smaller trauma from harsh words spoken in the classroom, the playground, at home or work—words that cause us to play safe rather than dare to take creative risks.

The Role of the Supportive or Critical Friend

Creating something new can be a lonely business. As we have discussed earlier, we are all subject to a myriad set of biases, blind spots and self-limiting beliefs. A mentor can play a useful role in encouraging perseverance, challenging those blind spots and engaging in a creative dialogue as described in Chapter 3. Who speaks in to your life—who can challenge you, who can spot the self-limiting beliefs?

Know who the right friends for the right occasions are. Know who to go to:

- when you need emotional support
- when you need critical feedback
- when you need encouragement
- when you need a kick up the backside

Be aware of when the right time for each form of support might arise. Enlisting critical feedback from a friend when the work is in its early stages of development is likely to elicit suggestions for improvement. But it might equally feel wounding and ultimately undermine continuation of the project, particularly if the feedback judges the work-in-progress against what a finished version of the project might look like.

However, critical feedback might be just what's required when you've looked at a creative piece of work so much that you can no longer take an effective appraising perspective on it. Feedback might be just the thing that helps complete the process.

A Caveat

There is a dark side to perseverance. As we've discussed, perseverance is key to seeing creative projects to their end, and to building resilience against naysayers and other forms of resistance.

However, ultimately, you as creator have little influence over any critical and commercial success your creation may have. You can increase your capacity to BE creative, and to complete your creative work. But you can't control how it will be received. Perseverance may lead ultimately to acceptance and success (and these won't happen without some amount of perseverance).

However, there are no guarantees of acceptance and success, no matter how much you persevere. So be mindful of how persevering on a particular project balances with other areas of life, and take action accordingly. If your creative pursuits are compromising your capacity to support yourself effectively in other areas of life, consider how you can make the necessary adjustments. Blindly continuing to persevere in the belief that critical acclaim and commercial success WILL come eventually is dangerous as there are no guarantees.

Chapter Summary and Conclusions

The practical realisation of creativity requires more than just a clever idea. It requires effort, perseverance, and self-belief to overcome the barriers. In seeking (and helping others) to make the most of our creative faculties, we should:

- Aim to develop grit, resilience, and mental toughness
- Consider how to deal with fear of failure and procrastination
- Reframe failure as a learning opportunity

- Use SMART goals, process goals, and sundowners to focus efforts and make progress
- Draw on the support of others when faced with challenges

Chapter 5
Techniques for Idea Generation

Introduction

Here is Edward Bear, coming downstairs now, bump, bump, bump, on the back of his head, behind Christopher Robin. It is, as far as he knows, the only way of coming downstairs, but sometimes he feels that there really is another way, if only he could stop bumping for a moment and think of it. – A. A. Milne (1)

We may well have a germ of an idea or have a problem we're aware we want to solve, but everyday life (bump, bump, bump) and everyday thinking always get in the way. So, what can we do to jolt us out of our everyday thinking—from convergent to divergent thinking—and encourage our creativity to flow?

In his book, *Creative Thinking and Brainstorming*, Geoffrey Rawlinson (2) suggests that there are six barriers to creative thinking:

- Self-imposed barriers that may be a result of too much analytic thinking. An example is the phenomenon known as 'imposter syndrome' where somebody believes, without evidence, that they are not good enough to be in the role they are in, and that sooner or later they will be found out. This limits their capacity to use their full potential.
- Established patterns of thinking which lead to believing that there is only one unique answer to a problem.
Consider this puzzle using the letters A-F:

<u>A E</u>
<u>BCD</u>

Where would the F go? Some might say it should go below the line because all those letters are consonants. Others might suggest that it should go above the line as those letters are all comprised of straight lines (whereas those below have curves).
- Conformity to significant others, and to expected norms. When the most senior people in a business meeting state an opinion or make a decision, it can be extremely challenging for more junior colleagues to express alternative opinions or advocate for other decisions. Instead, agreement is the easiest and most likely outcome.
- Not challenging the obvious. For example, solve the following: 1+1=? The obvious answer is 2, but other answers could be 11 or T. If you use the + and = you could even get a picture of a church with a cross on the top
- Evaluating the options too quickly because you have found one idea that initially appeals, or have rejected other ideas because they might be considered impractical or silly. This 'tunnel vision', where focus on one option inhibits an assessment of any alternatives, is very common.
- Fear of looking a fool or being reprimanded when identifying options that might bring ridicule or social rejection. The doctor in Wuhan who first raised the alarm about the coronavirus (and who subsequently died of the disease) was reprimanded for spreading false rumours.

The idea generation techniques that we discuss in this chapter, including Rawlinson's own 'restatement method' of brainstorming, can help to overcome these blocks.

As we've discussed, we take the view that creativity has many behavioural elements to it. Indeed, creativity is less about personal 'wiring' than about a motivated goal-oriented process of developing and expressing novel ideas for solving problems or satisfying needs. This can often take place through collaboration with others (3).

Knowledge and the deliberate use of learned cognitive strategies or techniques (i.e., ways of *thinking*), as well as the motivation to persevere with the process, are more significant to creative outcomes than differences in inherent individual traits or associative abilities (i.e., inherent abilities in divergent thinking).

As such, there's *a lot* that *anyone* can do to enhance their creative faculties and bring about creative ideas. There are a huge variety of techniques that can be employed to nudge us out of our patterns of habitual thinking, force us to take an unfamiliar perspective on a situation, help us to see ways to combine disparate elements in unexpected ways, or help us generate large quantities of ideas. These are also the techniques to use when it becomes tempting to give in to the little voice that says, "I'm just not creative."

However, there is of course a giant caveat. As we've already discussed, creativity has a number of critical ingredients—knowledge and expertise, intrinsic motivation, creative thinking, perseverance—and using idea generation techniques is not a panacea across the board. It is no simple substitute for the knowledge or expertise that is generally required in creativity but is best used in conjunction with them.

It may not help if you don't have the intrinsic motivation in place that is so important in creative practice. But it will help with the necessary creative thinking to approach problems from a different perspective, make novel combinations of existing ideas, and generate the quantity of ideas necessary to break free from convergent thinking and envision something new.

There is a huge number of idea generation techniques—as well as many variations on the general themes—and we certainly don't cover them all here. Not every technique will be appropriate in every situation. Since the concept of creativity covers such a broad range of circumstances, the most suitable techniques will vary accordingly.

What we cover is a range of techniques, grouped into a number of major categories, and explain how to use them effectively. We include some of the more popular and well-known techniques, such as brainstorming, mind-mapping and role-playing, as well as lesser-known variations on these themes. We aim to give you the major steps and important information so that you can use each technique effectively.

There are some overarching principles that appear through all techniques, which should be borne in mind.

1. Experimentation is a key part of the process. Creative ideas don't emerge without many other ideas preceding them. The more ideas you can generate, the more likely it is that these ideas will include some that could be considered creative. You also need to break free from the tried

and tested ideas that first come to mind out of habit. So, these techniques generally help you create a large number of ideas.

2. A lack of censorship or judgement is an important part of the process. Unfamiliar (and therefore novel) ideas can often be easily dismissed because they are out of the norm or they seem too far-fetched. This may well spell the premature end of a potentially creative idea. The assessment and judgement of ideas should come at a later stage in the process.

3. These techniques aim to encourage new connections and new perspectives. This may bring about a sense of unfamiliarity and perhaps discomfort. This is evidence that the process is working as it should, rather than a sentiment to shy away from.

4. A sense of playfulness is often at the heart of many creative processes. Many of these techniques try to harness this sense of playfulness to enhance our creative outcomes. Embrace the playfulness!

5. Many of these techniques encourage and facilitate collaboration. Whilst we tend to think—historically and culturally—of creativity in individual terms (e.g., Newton, Darwin, Einstein, Edison as lone warriors), much research points to creativity and innovation frequently arising in group situations—from team meetings to chance water-cooler conversations.

6. Do not be a slave to the tool! It is only an aid to producing ideas. If it is not doing that, change the technique or modify it in a way that works for you.

For many of us, like Pooh Bear, the opportunities for creativity and the ideas for improvement are often lurking semi-consciously, but they are unable to emerge because of the circumstances we find ourselves in. By applying idea generation techniques, we clear a space for proactive engagement in the creative thinking process.

Let's not forget, however, that knowledge/expertise and attitude/intrinsic motivation are important elements in the mix too; and having a creative idea is only the first step—you then have to DO something with it.

Chapter Summary

In this chapter, we identify and explain a selection of tools—grouped into twelve categories—that can be used by individuals or can deployed by a facilitator in group situations. Some tips on facilitation and the typical design of these events are outlined in Chapter 7.

1. Exercises to get into a creative mood and warm up the idea-generating 'muscle', or bring about a shift in perspective
   a. Boxes exercise
   b. The cake
   c. Uses for a paper clip
   d. Random word associations
   e. Six Thinking Hats (de Bono)
   f. Catch ball game
   g. 9 dots exercise
   h. Picture of George and Mrs Washington/old lady vs young lady
   i. Negative influences
   j. Worst ideas
2. Brainstorming
   a. Brainstorming principles
   b. Imaginary brainstorming
   c. Two level thinking
   d. Brain writing
   e. Restatement method of brainstorming
   f. Affinity diagram
   g. Visioning
   h. Scenario planning
3. Role-playing
4. Semantic techniques
   a. Analogy and metaphor
   b. Random words

c. Reversal
   d. Reframing and '5 whys'
   e. Story telling
5. Visual techniques
   a. Mind maps
   b. Fishbone diagrams
   c. Picture associations
   d. Drawing the problem
   e. Storyboarding
   f. Shield exercise
   g. Visual success map
6. Attribute combinations
   a. Morphology strips
   b. Ideas matrix
7. Serendipity
   a. Improvisation
   b. Randomness
   c. The unconscious and dreams
8. Research
9. Combined approaches
   a. TRIZ
   b. Roger von Oech's creative solutions
   c. TILMAG
   d. SCAMM, ESEAM, SCAMPER
10. Observational approaches
    a. Customer ideas
    b. Empathic design
    c. Modes of thinking
11. Selection of ideas
    a. Selection grid
    b. Multi-voting
    c. Impact analysis
12. Implementation
    a. 'Pebble in the shoe'
    b. Force field analysis

1. Exercises to Get into a Creative Mood, Warm Up the Idea-Generating 'Muscle' or Bring About a Shift in Perspective

a. Boxes Exercise

Try this exercise to get your creative juices flowing. It is an opportunity to explore your creative side in a fun way. It encourages you to break out of automatic and habitual thinking patterns and generate a large number of ideas, some more expected than others. The more unexpected the idea, the more it starts to meet the criteria for being deemed creative.

A variety of approaches are suggested to prompt a shift in perspective and demonstrate how numerous different ideas can be generated with different perspectives.

Exercise

Using each of the squares below, draw as many different ideas that come to mind as possible. There is no right or wrong, simply draw something—anything—in the square.

Now try filling the boxes again, using the following prompts and drawing the answers to the questions. This exercise encourages you to switch perspectives and generate new ideas as a result.

- Imagine that each box is:
    - An envelope. What is on the back?
    - The outline of a house. What is in it?
    - A gift. What is inside?
    - A television? What is on?
    - A picture frame. What is the picture of?
    - A huge box. What's inside the box?
- Draw a simple geometric shape inside a box. What is it now?
- Imagine that:
    - The box is something that you would find underwater. What is it now?
    - The box is red. What is it now?
    - The box is part of a car. What is it now?
    - You are a child. How would you see the box?
    - You are an accountant. What is it now?
    - The box contains something explosive. What is it?

Now randomly place your finger on one of the words listed below. How can this word be used to create a new and original way of looking at the box and prompt you to draw something new? Try several times with different words.

Diet Television Iceberg Flea Bed Ship Advertisement Hand Rock Chimney Hair Crown Glasses Train Japan Baby Clay Bear Violin Pizza Chemistry Snake Necklace River Unicycle

Now work as a group and see if more ideas emerge as you work together and bounce ideas off each other.

b. The Cake

This is a brainteaser that requires a break from assumptions to solve. It demonstrates how captured we can be by our mental assumptions, and how hard it can be to challenge those assumptions.

Exercise

It's a special occasion for you and your team, so you bring in a cake to celebrate. Your task is to cut the cake into 14 pieces, using only 4 straight slices of the knife. The pieces don't have to be of equal size.

Solution

Make 3 slices into the cake as in the diagram below. This will create 7 pieces of cake. Use your 4th slice to cut the cake in half horizontally, thus doubling the number of pieces to create 14 in total.

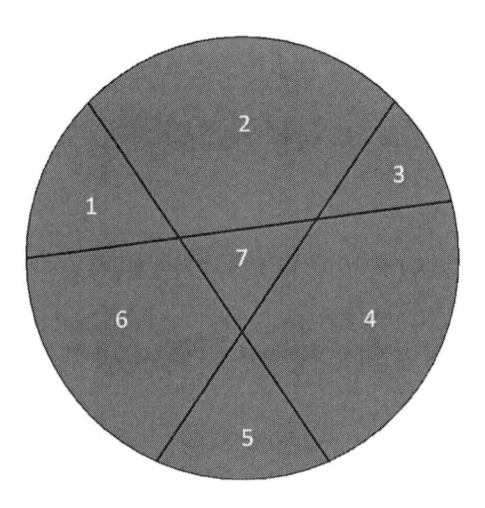

c. Uses for a Paper Clip

For this exercise, you can substitute the paper clip for any other mundane or simple household item. One common example is a brick. This exercise encourages people to focus on a large quantity of ideas, which means that they have to move past the predictable and expected ideas and generate more novel and unexpected ideas. This is often a very important part of the idea generation process, particularly brainstorming. Appendix A gives a list of uses to start with.

As a preamble to this exercise, you can describe the object in detail (or ask others to do so). This encourages close observation of the properties of the object, which may then lead to ideas that are associated with that particular property. For example, you might notice that the paper clip is metal, which could then lead to ideas based on its capacity to conduct electricity, or to heat up.

Exercise

Write down as many uses for a paper clip (or other object) as you can think of in 5 minutes (or 10 minutes).

Variation 1: Imagine you were someone different—from a different department in your company, or from a different country, or a Roman or Martian. How would they use the object?

Variation 2: Transport yourself to a different environment such as the Antarctic or the moon. What would you use the object for there?

d. Random Word Association

This is a good exercise to get a group warmed up. It can also be used as a warm-up segue into a creative discussion about the problem or issue that the group has met to address.

The process will most likely feel challenging and stiff and stilted at first, but after a while, people will find it easier to more readily contribute. As the exercise takes place, people's thinking processes shift; habitual self-censorship lessens as the focus switches from getting it 'right' to simply producing any word that meets the criteria.

Exercise

The group stands in a circle and one person starts the exercise by saying the first word that comes to mind (e.g., "yacht"). The next person in the circle must then say the first word that comes to mind beginning with the last letter of the previous word (i.e., a "t" in this instance, e.g., "tomato"). Keep the exercise going for several rounds of the whole group, until the group is sufficiently warmed up.

Once the group is warmed up, and has reduced the power of the self-censorship filter and grown in comfort with contributing thoughts and ideas as they come to mind, the group can then switch its focus to the issue or problem that needs to be addressed, and write down any associated words that come to mind. (You can do this on post-its, or as a collective list.) You can also use pictures (drawn or cut out of magazines) as an alternative to associated words.

An alternative approach is to use analogies and metaphors. Identify statements that are metaphors, e.g., financial *watchdog*, food *chain*, moral *bankruptcy*, hard as *nails*, sharp as a *knife*, and get the group to come up with similar phrases using analogies and metaphors that describe the issue or problem being addressed. (See Section 4 below for more ideas)

e. De Bono's Six Thinking Hats

Over the years, author and academic Edward de Bono (4) has contributed many ideas to stimulate creative thinking, such as his 'Six Thinking Hats' process. Each hat—which a team member wears, usually metaphorically—represents a different mode of thinking and interacting in a group.

A hat is a piece of clothing that you put on and take off quite easily. In many cultures, a hat is associated with the 'role' that someone is playing at a particular moment (e.g., a fireman's helmet, a policeman's hat, a bishop's mitre). It is also often metaphorically associated with thinking: "Put on your thinking hat."

The hats are:

- White—focus on facts and figures
- Red—focus on emotions, gut instinct
- Black—focus on why things cannot be done, playing the devil's advocate
- Yellow—focus on positive thinking about what can be done
- Green—focus on creativity and new ideas
- Blue—focus on facilitation of the thinking process

It gives permission to participants to play a different role or adopt a different thinking style. It can also be used to ensure that all team members are using the same approach at the same time, rather than working at cross-purposes to each other.

This works well as a warm up exercise because it can be a lot of fun for groups, but it can also work well as a reframing exercise, or as a tool for making sure group problem-solving works effectively.

f. Catch the Ball

This is an exercise designed to help a group to recognise that creative ideas can come from anyone—not just the boss or the creative department—and often require them to challenge assumptions or shift perspectives.

Exercise

The group stands in a circle, with one person holding a ball. The first person throws the ball to someone else, remembering who they threw it to. That person then throws it to someone else and so on until everyone in the group has received and thrown the ball. Then the ball is thrown back to the first person.

Then the group must circulate the ball around the group in the same order, only faster and without dropping the ball. If they succeed in doing this, they must do it again even faster. Sooner or later, someone drops the ball.

Facilitator prompt

Prompt the group to think about what they are doing, and whether there is a better way to increase the pace with which they pass the ball around the group. They might reflect that they were trying to work harder (increasing their speed) but not smarter (finding a different approach to passing the ball around the group).

Typically, someone comes up with an idea of getting closer to each other or rearranging the order in which they are standing so that everyone is passing the ball to the person next to them. Perhaps they might stand in such a way as to create a slope of hands so the ball rolls down the slope over everyone's hands. Finally, everyone might stand around the ball and touch the ball simultaneously.

Analysis

The final solution (i.e., everyone simultaneously touching the ball) comes about from experimentation and from pooling ideas from anyone. It also requires the group to challenge any assumptions that they might have about the rules of the game (e.g., that they are not allowed to change the order in which they are standing), and shift their perspective (it's not about *throwing* the ball, but making sure it *passes* from one person to another). These are important elements in the creative process.

g. The 9 Dots Exercise

This is a lateral thinking exercise that encourages people to 'think outside the box'—both figuratively and literally!

Exercise

Join all 9 dots using only 4 straight lines without lifting your pen off the page (i.e., the 4 lines are all joined).

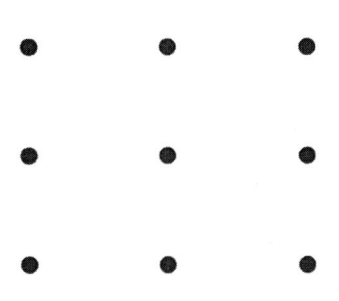

Solution

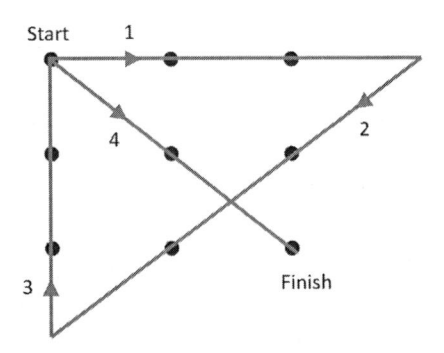

If someone already knows the answer, then challenge them to complete the exercise again using even less lines. One potential solution is to turn the piece of paper into a giant tube; in this scenario, one line could suffice. Or to (hypothetically) use a giant marker pen that is wide enough to cover all the dots with one line across them. These different approaches challenge people to realise how much we make assumptions about what is permissible and what isn't.

h. Picture of George and Mrs Washington/Old Lady vs Young Lady

Here is a picture of George Washington and his wife surveying troops.

And here is a picture of George Washington's face.

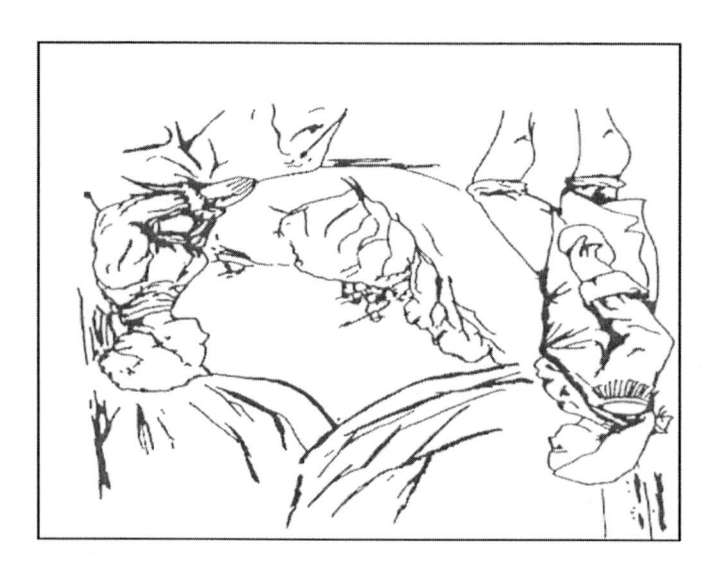

Or is it? In fact, it is of course the same picture, turned upside down. If you show the picture to half a group one way up (e.g., surveying the troops) and the other half of the group see the picture the other way up (e.g., Washington's face), and then discuss what everyone saw, it often transpires that each group only sees one picture and not the other.

This exercise challenges our understanding of what we see and how we interpret what we see. It demonstrates that a different perspective on the same 'information' will yield a different interpretation.

A variation of this exercise is to show a group the following picture and ask them whether they see an old lady or a young lady (or ask them what they see).

Some of the group will see the old lady looking down, while others will see the young lady turning her head away. Once you have seen one lady, it is very hard to see the other. This shows how hard it can be, once we have interpreted a situation in a particular way, to reinterpret it differently. In situations where creativity is required, we need to consider how we can break out of our habitual habits and modes of interpretation and look for new ways to interpret a situation. Certain idea generation techniques will help with this!

i. Negative Influences

Exercise

Take a sheet of paper and write down all the negative influences or thoughts that might inhibit you in the creativity session that is about to follow. Then tear off each influence or thought individually, screw it up, and throw it away. For each one, say "I will not let this affect me today".

Analysis

So many aspects of the creativity process are undermined by the barriers we create for ourselves. Negative thinking, self-doubt, anxiety, stress, etc. can all inhibit creative thinking. By explicitly naming these influences and then pledging to be unaffected by them, we give ourselves a better chance of preventing them from derailing our creative process.

j. Worst Ideas

Exercise

Ask people to brainstorm the worst ideas they have ever come across, from any field. They can be examples of bad, stupid, illegal or just plain silly behaviour. For example, it could be the man who recently came a cropper trying to do handstands on the guard rail of a ferry, and promptly fell overboard and drowned. There is often amusement and laughter, and a free-flow of contributions as energy is re-generated.

The next step is to see whether these ideas could actually be turned into something useful.

Analysis

This is an exercise in brainstorming and perspective-shifting. It is very useful to re-energise a group by bringing some humour and absurdity into proceedings.

2. Brainstorming

There are a huge number of approaches to brainstorming, and many of the techniques that we have placed in the other categories also incorporate brainstorming elements in them. Its main purpose is to generate as many ideas as possible, with the goal that there are some creative ideas to be found amongst them.

Edward de Bono called the mind a 'patterning system', gravitating towards familiar patterns. By aiming for a large quantity of ideas, we force ourselves to go beyond the habitual and expected ideas that first come to mind, ideas that are usually based on past experiences, normal conventions and common assumptions.

The ideas generated later in the process tend to be more unexpected, often based on shifts of perspective or novel combinations, and tend to yield creative ideas amongst them.

We start this section by describing the basic features of brainstorming, and then we describe a number of different variations on the basic premise.

Note: there has been a suggestion in recent years that the term 'brainstorming' is not politically correct as it may cause offence to some people. We apologise if this is the case with any reader; we do not intend to cause offence but are using the term as it is so commonly used that other terms might not be well understood.

a. Brainstorming Principles

The purpose of brainstorming is to generate as many ideas as possible in a relatively short period of time. There are some key principles to ensure that this process happens as effectively as possible.

1. Define the problem

Make sure you know exactly what it is that you are brainstorming. It can be hard to come up with ideas if you are vague about what those ideas are supposed to achieve. For example, a brainstorm on one's future may be too broad and vague, whereas a brainstorm on one's future *career* might be more useful.

Some people have suggested that one should start by defining the problem, and then leave some time (e.g., 2 hours) before commencing the brainstorming process to generate ideas. They argue that this gives the mind some time to subconsciously work on the issue and prime it with ideas. However, others argue that the ideas that may be consciously generated in this time period may be lost as they are not being recorded as part of the brainstorm.

2. Generate as many ideas as possible

Quantity of ideas is vital. Don't dwell on discussion of certain points as you generate them, simply note them down and move on or build on them to find new ideas. There will be an opportunity later in the process to scrutinise the ideas more closely.

It can sometimes be useful to set a stretching goal in terms of numbers of ideas. For example—50 ideas out of a 15-minute brainstorm. This forces people to focus on the quantity of ideas without over-thinking or censoring the ideas that arise.

3. Withhold judgement of ideas

All ideas are acceptable and should be written down or recorded. No criticism should be offered—either of others' ideas within a group setting, or of one's own ideas by way of self-censorship. There should also be no implicit censoring of ideas such as not writing ideas down.

Even if ideas seem mundane, or downright ridiculous, record them nonetheless. They may well lead to more appropriate ideas. Say the ideas as you think of them—they don't have to be explained or justified in any way (at this point). Be encouraging of others, however zany or ridiculous the suggestion seems. Accept that each suggestion is legitimate at the time it is made.

Duplicate ideas are acceptable; even if the idea has already been offered, write it down as it keeps people engaged in the process.

4. Build on (each other's) ideas

Use the ideas as springboards for further ideas or try combining ideas that have been generated to create new ideas. Cross-fertilise and use others' ideas as a springboard for your own.

Sometimes, an idea that has been expressed might not appear to have merit or might be so far-fetched as to be ridiculous. But these ideas can be very useful as they can often spark other ideas in people, which can lead to creative solutions.

The British scientist Magnus Pyke told the story that in the Second World War, Britain was under intense pressure from the German bombing raids, and, with aviation fuel in short supply, the British Royal Air Force had a dwindling capacity to challenge and repel the bombers and conduct bombing raids of their own.

During a brainstorm, someone suggested that since aviation fuel was in such short supply, both sides should just bomb themselves rather than fly all that way to bomb each other—a ridiculous idea. However, out of this idea emerged a plan to paint factory rooves with bomb craters.

The enemy bombers saw these 'bombed out' rooves during raids, thought that there was no further need to bomb that location and moved on. It saved many factories that were critical to the war effort from destruction.

The idea of Neighbourhood Watch is reputed to have come from the Los Angeles Police Department. Faced with rising crime and limited resources, they brainstormed ideas for new approaches, and someone suggested that if they recruited all the criminals into the police force, they wouldn't have to catch them.

At first, this idea seemed ridiculous, but it evolved into the practical solution of neighbourhood watch schemes, whereby local coordinators kept the local population informed of crime risks, reported things that were suspicious and watched out for each other's property.

5. Encourage crazy and exaggerated ideas

Encourage flights of fancy, a belief in the impossible, and the downright weird, crazy, profane or even illegal. The aim is to generate a large number of ideas, and also ideas that are novel and unexpected (i.e., creative). If we are bound by convention and politeness, as well as the apparent realms of possibility, then we restrict and undermine our capacity to generate these novel and unexpected ideas that are deemed to be creative. Once the brain can start free-wheeling, free from constraint, then the ideas can start to flow.

Some other pointers

Write every idea down, on paper, on post-it notes, on a flipchart. This can be done collectively with one designated scribe, or anyone can write down an idea as they have it. One group each individually wrote their ideas down on paper planes and threw them to each other.

Don't end the brainstorm too early. It can be a good idea to agree on a timeframe before starting. It is common that a brainstorm will slow down or even grind to a halt after an initial flurry of ideas. But these initial ideas are usually the predictable and habitual ideas and responses; the slow-down is the point at which these predictable ideas have run dry. This is the opportunity to push ahead with the brainstorm and begin to elicit the more unusual and creative ideas.

However, don't force the brainstorm to go on too long. If energy is flagging or does not pick up again from a slow-down, then bring the session to an end, or set a short time limit e.g., another three minutes. Alternatively, you can re-energise the group by reviewing the ideas so far and building on them further.

In a group brainstorm, it is important that everyone participates and contributes. This requires encouragement and non-judgement from the group. There are two general approaches to brainstorming, structured and unstructured, and the best approach will depend on the extent to which everyone is able to contribute equally to the process.

In a structured brainstorm, each person contributes an idea in turn or passes until the next round. The session ends when everyone passes. This method gives everyone a chance to contribute but can put pressure on members to 'perform' on cue. The facilitator needs to establish clearly that it is OK to pass, and that it is possible to contribute more ideas after an initial pass.

However, it does ensure that everyone gets a turn to speak up, and this can be very useful in group situations where there are individuals who dominate proceedings, individuals who are too timid to speak up without permission, or cultural conditions that make it challenging for individuals lower in a company hierarchy to contribute when there are more senior individuals present.

In an unstructured brainstorm, people just offer ideas as they occur. This may lead to stronger group interaction and to a greater sense of group energy. However, it also runs the risk of being dominated by a few strong group members. It can also be hard for a single facilitator to write down all the ideas as they may come too fast. If there are two people to scribe it is often easier.

When the brainstorm is over, the group must evaluate the suggestions—of which there are hopefully a large number—and decide which ones to explore further or experiment with. There are a number of ways of choosing the best idea(s). Voting is one approach, particularly if a single 'best' idea is required.

Everyone picks their top 5 choices, and the idea with the most votes is chosen. Alternatively, if ideas have been captured on a flipchart, people can be given a number of stickers to allocate against their favourite ideas on the flipchart (see multi-voting in section 11).

Alternative approaches

Group members can contribute their ideas by writing each idea on a post-it note and sticking it to the wall or a flipchart. Once the brainstorm has ended, the post-its can then be grouped according to theme. Or they can be placed on an impact analysis chart like the one discussed in section 11. This sorts out the ideas in terms of their level of impact and how easily they can be achieved. Ideally, look for the masterstrokes that have a high impact in the problem area and are also easy to achieve.

A brainstorm with post-its can be approached in either a structured or unstructured way, and ideas should be called out as they are written down, so that everyone else has the opportunity to build on them or bounce off them. A facilitator can collect the notes as people write them down and call them out, or people can stick them to the wall or flipchart themselves.

b. Imaginary Brainstorming

The challenge that all idea generation techniques are trying to overcome is our natural tendency to think in the familiar ways and patterns with which we are comfortable. Imaginary brainstorming deliberately sets out to look at a partially related but very silly problem. This encourages the group to safely leave their comfort zone by giving them permission to be silly. If the whole premise of the issue being brainstormed is ridiculous, then it clearly doesn't matter how ridiculous the suggested ideas are.

For example, suppose we were trying to think of a way of speeding up the time required for maintenance of a machine. We might apply this imaginary brainstorming technique and start by thinking of ways of speeding up a night out, dieting, painting the house, dating, going to the moon, etc. Then we can pause and see if any ideas can be transferred to the real problem.

It can be useful to alternate between brainstorming 'sensible' and 'silly' ideas and scenarios.

c. Two Level Thinking

We tend to think at either 'incident' level or 'concept' level. Incident level thinking entails thinking about a specific situation. Concept level thinking entails thinking about a general or theoretical concept.

Linking the two can help to increase the range of ideas.

Process

- Carry out a standard brainstorm for the problem in question (incident level thinking)
- Cluster the ideas into themes and concepts, and create an affinity diagram (see the detailed explanation later in this section)
- Take each theme or concept that has emerged, and brainstorm further ideas related to the theme or concept (concept level thinking)
- See if any of the ideas generated during the concept level brainstorm can be applied back to the original (incident level) issue

For example, if a group were brainstorming ideas about improving a product, one of the themes that might emerge might be around speed (of manufacture, of delivery, etc.). The group could then brainstorm ideas around the concept of speed, and of doing things quickly. They could then apply these ideas back to the specific situation of their product.

Another example might be a group who are brainstorming ideas for the promotion of a new product. A key theme that emerges is of getting people's attention, so the concept-level brainstorm focusses on all the different ways the group can think of to get people's attention (e.g., jump in front of them, make a loud noise, grab them by the arm, go up to them with a puppy!). The group can then apply all these ideas for getting someone's attention back to the issue of promotional strategies for the new product.

It can be useful to conduct this brainstorm using post-it notes; this makes it much easier to group the ideas by themes and concepts.

d. Brain Writing

This is a silent method of brainstorming. Each person in a group takes a sheet of paper, and writes the agreed issue or problem at the top of the page. They then write 3 ideas on the paper and pass it on to the next person in the group. Beneath these ideas the next person writes down another 3 ideas that have been triggered by the first 3 ideas, and passes the paper on. This continues until each piece of paper arrives back at its originator.

An alternative approach is to have a flip chart on the wall for each participant. Each person starts with a few minutes' 'brain dump' of their ideas before moving on to the next flipchart and seeing if they can add anything there.

As a variation, the exercise can start with a different problem at the top of each piece of paper.

This approach can be very useful when deeper thought is required for the ideas that are suggested, or where a group comprises more introverted individuals who don't like speaking up in groups.

e. The 'Restatement Method' of Brainstorming

This method was first developed by Geoffrey Rawlinson (5).

An initial brainstorm seeks to identify as many different ways as possible to restate or reframe the initial problem. For example, an initial problem of increasing sales could be restated as a problem of making customers buy more, or of attracting more customers, or of making past customers return with an offer they can't refuse, etc.

Each of these restatements can then be brainstormed individually to come up with multiple ideas, potentially unearthing solutions that would not have emerged through simply brainstorming the problem as originally stated.

f. Affinity Diagram

This is a useful tool for organising disparate ideas into meaningful clusters after a brainstorm. It is group-energising, fun and a very effective way to focus on broad themes, while maintaining all the individual 'pieces' for further group discussion.

Brainstorm the issue in question and write each idea that is suggested on a separate post-it note. When the brainstorm has finished, related post-its can be grouped together. An easy way to do this is to stick them on a blank wall. One post-it that best captures the essence of each cluster can be chosen; this way, the key themes related to the issue in question are identified.

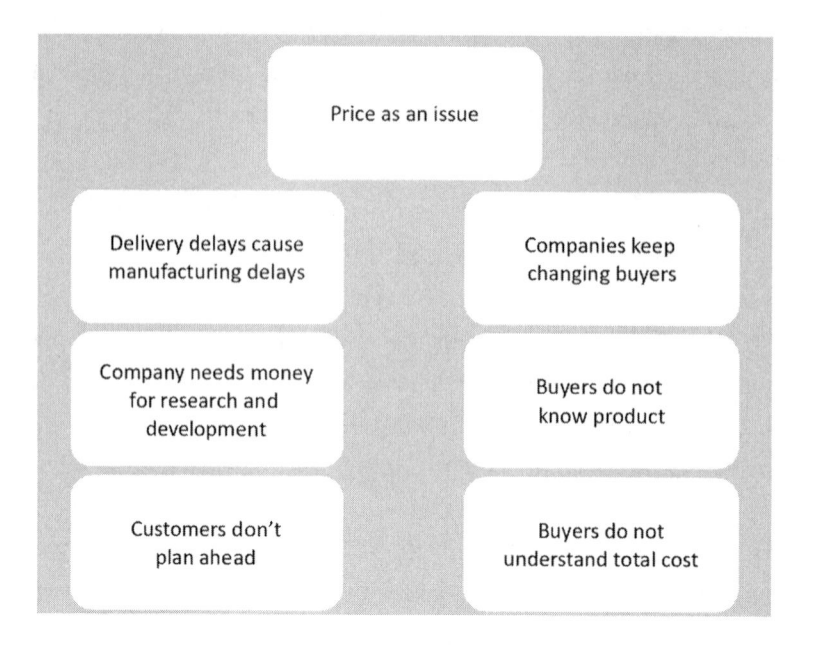

```
                    Price as an issue

    Delivery delays cause            Companies keep
    manufacturing delays             changing buyers

    Company needs money
       for research and              Buyers do not
         development                  know product

      Customers don't                 Buyers do not
        plan ahead                understand total cost
```

g. Visioning

1. Use visioning to articulate the future state and steps to get there
2. Use visioning to create a mood or energy for action

'The fish never feels the water', so the maxim goes, meaning that when you are in an environment that is all you have ever known, it is difficult to conceive of anything outside of this environment. Likewise, it can be hard to envision a future state that is different from the one you are currently in.

"Prediction is very difficult, especially about the future"—atomic scientist Niels Bohr.

"Guitar groups are on their way out"—Dick Rowe, Decca Records executive (when turning down the Beatles).

Giving some thought to the future that you want to achieve can be very helpful in coming up with ideas about what to do next. If you don't know the future state you're aiming for, it's very hard to know what to do in the present that will take you in the right direction.

This failure to clearly envision and articulate future goals is the cause of stagnation in many businesses and individual lives.

So how do you peer in to the future and free yourself from the mindset of the past? Here are some questions you might start by asking yourself.

- What might you do if anything was possible?
- What might you do if time and money were no object?
- Imagine a time in the future when the problem you are facing has been solved. Or the product you are developing has been successfully launched and has become wildly popular. How does it feel? How does this future situation look? Where are you? What are you doing?

As you start to articulate a vision of the ideal future state, you can then start to identify the steps you need to take to get there.

This 'blue-sky thinking' technique is very useful when you don't know what to do next or feel stuck and stagnant. When we are stuck in the treadmill of day-to-day life, often rushing around from one commitment to another, we just plough blindly ahead without much thought for where we are going.

Thinking forward helps to shift the focus and builds energy to move forward. It can energise us and get us excited about future possibilities. Our imaginations can produce images that take us beyond our conscious selves.

Here are some exercises to help you consider the future.

Exercise 1

Stand up and point your finger straight out in front of you. Let your eye follow your arm out to your fingertip. Keeping your feet planted in the same spot, rotate your trunk from the waist and gently twist as far as you comfortably can. Note the point that your finger reaches. Return to the starting point. Now close your eyes and imagine yourself doing the same thing, but this time going beyond the point that you previously reached.

Now allow your body to follow your mind. When you open your eyes, you will find that you have gone further than you did a moment ago (6). This is a nice analogy for how imagining a certain future state can help us achieve it, even if it initially seems to be beyond our capabilities.

Exercise 2: What do I want to be said about me?

Imagine you are at a party. As you enter you see all your friends and family, work colleagues and acquaintances. It is a party for you. They are there to celebrate your contribution to their lives. What would you like them to say?

Imagine that four people have decided to speak. The first is from your family—maybe your partner or child. The second is from your work—maybe a boss or a direct report. The third is from your community, church or other organisation in which you have served. The fourth is from your group of friends.

- What kind of person will they say you are?
- What contribution will they say you have made in your work and to your community, family and friends?
- What difference will they say you make to their lives?
- What will people thank you for?

After you have thought about it for some time (and perhaps made some notes), consider whether there is anything they won't say that you would have liked them to say, or—worse—anything they might say that you would wish them not to say. And if there are any such things, consider further how you might change yourself and your life so that the speakers would say only what you would be pleased to hear.

Now make a list of actions you will need to take to make this possible.

Exercise 3: Write the front page of a periodical

Imagine you opened a periodical written about your organisation in two years' time. What would it be saying? What would be the headline on the front page? Make notes about the stories you would like to see there. What are the achievements that are described, by whom were they made, and what did they do to get there? What will the stories be saying about *you*? What do you need to do to make sure these outcomes are achieved?

If this seems unfeasible, consider instead stories about your organisation that you would *not* like to see in two years' time. What should happen to ensure that such stories do *not* appear?

This can be done as an individual or group exercise.

Exercise 4: My perfect working day

Imagine that you could construct a perfect working day. What would it be like?

Start by imagining yourself getting up—where would you be? What time would it be? Where would you be going to work? Imagine yourself starting work. What would you be doing? With whom? In what environment? What would be going on? Picture all the activities in a day that would give you satisfaction.

You may want to contrast it with a terrible day—one that you would like to avoid at all costs. What did you learn from this? Do you understand any better what it is that you want in any future job or lifestyle?

Visioning can also be a good tool to get us into a frame of mind for action.

As before, imagine a time in the future when the problem you are facing has been solved. Or the product you are developing has been successfully launched and has become wildly popular. Imagine the party to celebrate and the accolades heaped on you for solving the problem or launching the product. How does it feel? Good? Now you are in a good frame of mind to get down to work.

If you are procrastinating, imagine the consequences of not meeting the deadline.

Take your mind away to a place where you have had your best ideas. Ideally, could you go there physically? Can you at least change your current surroundings to recreate as many elements of this inspiring place as possible: lighting, music, the room you are in, which way you are facing, with or without people?

Remember the personal conditions for creativity that we discussed in Chapter 3. How can you put all those elements that you personally identified in place?

h. Scenario Planning

The petroleum company Shell is widely credited with propounding the technique of scenario planning as a way of overcoming the natural tendency to over-emphasise extrapolation of the past. The underlying assumption with many forecasting tools is that the future is already lying there, hidden and waiting to be discovered, the inevitable result of past events and patterns, in the way that the natural sciences assume that nature is there to be discovered. But this approach of looking at alternative scenarios is based on a different philosophical perspective.

Scenario planning assumes that the future is *not* out there waiting to be discovered, but is a fluid product of the myriad decisions that we and others around the world make today, with numerous potential outcomes. If we examine the different ways the future might unfold, based on these influencing factors, we can begin to determine actions that will move towards a more desirable future and away from a less desirable one.

The key steps are:

- Determine the scope and timeframe of the project.
- Identify the current assumptions and mindset of the decision makers. Clearly, if you come up with scenarios that go against current thinking, they will need to be more convincingly argued.
- Create alternative plausible scenarios of how the future may evolve, identifying key assumptions, signposts that might indicate which scenarios are more likely to emerge, and critical decision points. There are some generic scenarios that are worth considering as part of a scenario plan: doing the same in a changing world, the best possible case, the worst possible case, catastrophe, and technological breakthrough that enables a new way of working.

The process of making decisions, such as deciding which technology to back or location to move to, is fraught with dilemmas. The longer that options can be kept open, the greater the chance of making an optimal decision, but the greater the chance also of missing some competitive advantage.

- In examining the assumptions, consider what PESTEL (Political, Economic, Social, Technological, Environmental and Legal) factors are driving which scenarios. You may wish to focus particularly on those factors that have high impact and are confidently predicted.
- Test the impact of key assumptions and other variables on the different scenarios.
- Develop action plans based upon:
  o The most robust solutions across all scenarios.
  o Strategies that leave options open until the most likely/desirable scenario is clear without foregoing competitive advantage.

- o Critical actions that promote the chance of the most desirable outcome.

3. Role-Playing

Role-playing is a mechanism that enables you to take a different perspective, according to the role you are playing. By playing different roles, therefore, you can find yourself looking at issues and problems in different ways (essentially, this is a mechanism that can help you to reframe a problem or situation, as we will discuss shortly).

This process may result in defining the problem in a particular way and may well influence the solutions that you find. How you define the problem will also be influenced by the knowledge and information that you have already acquired (7).

A professor asked someone in the audience of a lecture she was giving whether she might shake their hand. Upon doing so, she asked the audience to imagine themselves as epidemiologists. "What is going on here?" she asked. The resulting discussion focussed on topics such as the transmission of micro-organisms and the spread of disease.

"Imagine now that you are anthropologists," she said, and the discussion turned to greetings and other customs. The professor then explained that the way we define our frame of reference—in this case, denoted by the roles that the audience was asked to assume—will determine the paradigm and knowledge base with which we assess the situation, which in turn will determine the questions we ask, the solutions we get and the limitations we will experience.

Finally, she asked the audience to imagine that they were nuclear scientists. "What is going on now?" There was silence, and the professor explained that sometimes, the paradigm of thinking is just not suitable to appraise the issue at all.

Role-playing has been used in training courses and simulations for many years. As a training tool, its aim is to place individuals in situations that are as close to what they might experience in real life as possible. This allows them to learn new skills experientially. It is sometimes criticised for not being realistic enough but can nonetheless be a very beneficial approach.

As a tool for stimulating creativity, it can be used to place people in novel scenarios in order to shift their perspectives, or expose them to new experiences, and see what happens. People can be asked to participate in a scenario, or to brainstorm whilst assuming new roles.

For example, a space exploration programme in Moscow experimented with role-playing in preparation for a mission to Mars. The mission's crew were kept in a simulated space craft for the estimated timeframe of the journey to the red planet. They then undertook certain tasks and challenges, such as dealing with emergencies like a broken leg, to see what solutions they would come up with.

Role-playing not only helps people to shift their perspective, but also to lose some of their inhibitions. If they are playing a character or role, then any particular behaviour can be attributed to that character, rather than to the individuals themselves. This way, the individual can escape any fear of censure or judgement.

The musician Brian Eno recalls the effect of role-playing among his family, and his use of the technique for inspiration in the recording studio: "At Christmas…the whole family play quite elaborate games which allow normally retiring people to become enormously extrovert and funny. [The games] free you from being yourself; you are 'allowed' forms of behaviour that otherwise would be gratuitous, embarrassing or completely irrational. Accordingly, I came up with these role-playing games for musicians." (8) Eno would then give his musicians prescribed roles in the studio, in order to encourage them to think about their music-making in novel ways.

In one scenario, roles include a futuristic musician, "who is in one of the 'Neo science' bands and plays in an underground club in the Afro-Chinese ghetto in Osaka; a member of an early twenty-first century 'Art and language' band, and a player in one of the first New Afrotech bands which play in a suburb of Lagos, the new Silicon Valley, where a number of high-tech companies are located."

Another approach to role-playing might be to view an issue from the perspective of each stakeholder in the situation. For example, an organisation might view a particular issue from the perspective of their customers, their board members, their community, their employees, their senior executives, their competitors, their competitors' customers…and so on. They can brainstorm solutions from each perspective.

When you adopt a role, you can ask yourself the following questions.

How would this person:

- Think
- Perceive the problem
- Explain the problem
- Perceive the ideal solution
- Use particular tools (and which tools)
- Take action

The country musician Steve Earle speaks of how, when faced with creative and career decisions early in his career, he would ask himself, "What would Neil Young do?" This would allow him to take a slightly different perspective from his own, and to tap into a mindset that has already brought about musical success and acclaim. Similarly, the British pop singer Leona Lewis and her team would consider what Whitney Houston would do in certain situations.

The TV show *Undercover CEO* saw CEOs work in a role in the lower rungs of their organisation—often without their peers at this lower level knowing who they were. This helped them to understand what it was like to work at all levels of the organisation, and also gave them a fresh perspective on their business, often yielding novel ideas on how to conduct business.

A training exercise in a company took a team of senior people and set them up as a competitor. They asked the senior team to look at their company as if they were the competitor with the aim of attacking weak spots where they could find them and coming up with more attractive products and services.

Very quickly, under the mantle of the competitor, the team was able to expose weaknesses that they had previously been reluctant to admit. The exercise also gave them the opportunity to articulate the brand values of their company, and of other competitors, that they admired. They then considered how they could apply these insights and ideas to their own products.

As well as changing roles, changing your location or environment, or the people with whom you are collaborating, can also bring about a shift of perspective.

4. Semantic Techniques

We could do worse than to follow Rudyard Kipling's (9) lead when he wrote:

I keep six honest serving men
They taught me all I knew
Their names are: what and why and when
And how and where and who.

a. Analogy and Metaphor

What do the following phrases have in common?

- A financial watchdog
- An operational bottleneck
- A communications network
- The flow of time
- The food chain
- A leap of thought
- A frame of reference
- Moral bankruptcy

They are all examples of metaphor. We can use the principles of analogy and metaphor in our creative approach to a problem. One approach might be to take an established analogy or metaphor that pertains to our problem and consider solutions that make sense for a literal reading of the metaphor/analogy.

For example, if we are considering morality, we could use the metaphor moral bankruptcy as our starting point, and then consider how one can get out of bankruptcy. Could you file for bankruptcy protection regarding morality? Could someone 'lend' you some morality to bring you out of bankruptcy?

We can take familiar metaphors and reverse them or alter them, for example, a financial watchdog versus a financial lapdog. We can also create our own unique metaphors or analogies for the issue, and then use them as a springboard for creative solutions. For example:

Traditional metaphor	Different metaphor	Implication
Financial watchdog	Financial ostrich	From scrutiny to ignoring/head in the sand
Operational bottleneck	Operational lock gate	From blockage to control
Communication network	Communication web	From a structured arrangement to a more random set of connections
Food chain	Food for thought	From linear hierarchy to reflection
Leap of faith	Acrobatic display of faith	From an unpredictable and precarious situation to a more controlled and skilled scenario
Frame of reference	Space of reference	From the boundaries to the opportunities within the boundaries
Morally bankrupt	Morally frugal	From condemnation to a recognition of relative values and multiple truths

Another approach is to look for analogies out and about in nature, or to look at a condensed part of the situation as a microcosm of the entire situation at hand. Fractals were discovered from the process of looking at the way in which part of a leaf resembles the whole of the leaf.

Different languages can yield new colloquial metaphors or analogies to work with. The French colloquialism for the English phrase 'it's raining cats and dogs' is 'it's falling cords' in reference to the straight unbroken strings of rain falling from the sky.

How might the idea of rain as a cord from the sky yield some new ideas? Likewise, how might the idea of falling rain being cats and dogs tumbling from the sky provide fresh ideas for non-English speakers?

Finally, it can help to create a problem statement for the issue at hand, and then consider any synonyms for words in this statement, perhaps with the help of a thesaurus. This exercise, called 'opportunity redefinition', can help to spark new ideas for solutions and approaches.

- Write a statement that clearly defines the objective (the objective is usually around finding a novel solution to a problem).
- Take the key word(s) within this statement and generate alternatives to these words. You might use who, why, what, where, when, how questions to do this.
- Then substitute these new alternative words into the objective statement. The new statements might not provide the answer in themselves, but might provoke further creative discussion and insight that yield solutions.

b. Random Words

Choose words (or pictures) randomly and consider how they could apply to the issue at hand. You can randomly open a dictionary or a book to locate the random words or use an online random word generator. Don't skip a word—it's often the words that appear most unrelated to the issue at hand that become the catalyst for the most creative solutions.

You can also try random juxtaposition—take two words randomly (e.g., one noun and one adjective) and put them together. What ideas related to your issue do they inspire?

A further approach might be to methodically cycle through each letter of the alphabet and see what solutions beginning with that letter spring to mind, for example, all solutions beginning with the letter A. This could be a useful approach to brainstorming a one-word slogan or tagline, or indeed a brand name.

c. Reversal

A simple semantic approach is to reverse the issue at hand, or the key questions being asked in order to solve the problem.

For example, if the problem is: 'How do we make our product memorable?' we could ask instead: 'How do we make our product forgettable?'

Instead of: 'How do we increase sales?' we could ask, 'How do we make sales decrease?'

This approach seeks to identify what are the key make-or-break factors in the success or failure of the issue.

By looking at the opposite scenario, we can also potentially uncover some 'unknown unknowns'. The Sufi philosopher Nasrudin once posed the question:

If two astronomers met and one only lived in the day and the other only at night, what would they say to each other?

This illustrates how we are bound up in the perspectives we take, and that seeking to explore an opposite perspective can yield new information. This idea of inspecting opposite sides of an issue stands as one of the founding principles of democracy in ancient Greece, subsequently codified by the German philosopher Georg Hegel as the Hegelian Dialectic: a proposal is made (the thesis), that is then questioned and opposed (the antithesis), and so, through debate, an agreed conclusion is reached (the synthesis). This is the foundation of Western thought and the scientific method.

When someone is being considered for sainthood in the Catholic Church, a 'devil's advocate' is appointed to challenge and argue against the canonisation. This process allows for the proposal to be examined from both sides and prevents groupthink and the mass being swept along in an emotional tide.

Exercise:

1. Take the issue or problem question
2. Reverse it
3. Brainstorm solutions to the reversed question (e.g., how to decrease sales)
4. Flip those solutions back around so they now pertain to the original issue or problem

OR

1. After a brainstorm, once several viable ideas have been identified, reverse these ideas and examine the consequences of executing them in reverse
2. Take any further insights that this process elicits back to the original problem

d. Reframing

Like reversal, reframing gives us the opportunity to examine the problem from a different respective, in the hope of revealing new insights and ideas.

There is a familiar way of describing two attitudes to life: the glass is half empty or the glass is half full. Both statements are true but reveal two different perspectives on the glass of water.

In a brainstorming situation, we can generate a number of different perspectives to take (or different roles, similar to role-playing—see the Brian Eno example above), and then examine the problem from each perspective. In a group situation, everyone can cycle through all the different perspectives together, or different perspectives can be assigned to different individuals for the duration of the brainstorm.

For example, we might look at an issue from the position of a customer, or of the financial regulator. Or we might look at an issue from the perspective of a Martian or a south sea islander that has not experienced the modern world.

The tool company Black and Decker once received a letter from a lady in Pontypridd (in Wales), asking for someone from the company to come and mow her lawn. She'd just bought a Black and Decker lawnmower, she said, but couldn't understand the instructions. (Black and Decker were selling tens of thousands of lawnmowers each year at the time.)

Subsequently, Black and Decker introduced a test on each new product, based on the question: 'How would the lady from Pontypridd get on?' to establish that anyone could successfully use the product. This question allowed them to shift their perspective from expert in the field to non-expert consumer.

Experts sometimes make assumptions that 'surely everyone knows' certain facts and processes. However, this is not necessarily the case, and considering a situation from a non-expert perspective can unearth these assumptions and prevent any issues they might cause. This is pertinent when product instructions are being written. They should always be checked by someone who has no prior knowledge of the product.

Another approach is to employ the restatement method, as described earlier, and explore whether there is another way of defining or stating the issue at hand. The way the problem is defined or articulated can dictate the kind of solutions that are suggested. By reframing the problem in this way, new solutions and insights may become evident.

Einstein supposedly stated that if he had one hour to solve a problem, he would spend 55 minutes defining the problem and determining the 'proper question', and 5 minutes solving it—once he had ascertained the proper question, it would be easy to solve. This demonstrates the extent to which the solutions are bound up in the way a problem is defined and framed.

For example, if we were experiencing trouble finding a street parking spot for our car every night, we might conclude that there are not enough street parking spots. Or we might frame the problem as due to too many cars on the road. However we frame the problem will dictate the type of solutions we consider (more parking vs less cars). Or we might have a problem of not selling enough products and reframe it as a problem of customers preferring our product less than that of our competitors.

How do you find what Einstein called the 'proper' problem? One approach is the '5 whys'. Ask why a problem situation has occurred and keep asking the question 'why' up to five times, or until you have found the root cause, or the solution becomes apparent.

It often takes a bit of digging to ascertain what the 'proper' problem really is; the 'proper' underlying problem might not be the problem that is initially evident. The problem that is initially presented might just be the elements of the problem that are first seen or experienced. It might be that the real problem is politically difficult to talk about and so only the more socially acceptable aspects of the problem are presented.

People often complain in organisations of 'poor communications'. The real problem might be that they feel left out of the communication loop, or that there is a lack of transparency, or that their specific needs are not being heard.

Another approach is to identify and list all the assumptions you are making about the problem you are tackling. Then you can rigorously question these assumptions until you have uncovered any incorrect assumptions. This process can help you identify new insights and solutions as a result.

We can reframe our perspectives by examining our modes of thinking—we all think in different ways at different times. Below is a list of some of the more common ways we think. The aim of this approach is to encourage a group to think more widely about a subject and in particular to look at the subject from different angles.

The facilitator steers the group to all consider the problem using a particular mode of thinking, and then to change to a different mode. A variation is to assign different modes to different participants and encourage debate. This is similar to the 'Six Thinking Hats' process we discussed earlier.

Creative thinking	Practical/factual thinking
Detail thinking	Big picture thinking
Empathetic thinking	Selfish thinking
Positive thinking	Negative thinking
Emotional thinking	Logical thinking

e. Story Telling

At workshops and other events, stories can illustrate or emphasise some aspect of organisational behaviour or culture that you espouse and aspire towards. In particular, stories that challenge an organisation's rituals, myths, ceremonies and symbols can give people confidence to challenge the status quo. Here are some examples.

Story 1

Once upon a time, a manager of a car plant managed to save the plant from business disaster through inspirational leadership, clever marketing and tight control of costs. The employees were relieved and thankful, and they clubbed together to present him with a car.

They worked on this car to make sure it was perfect—significantly more than they usually did on the cars on the assembly line—and when the day of the presentation arrived the whole plant was assembled. The manager stood up and made a great speech about commitment, hard work, integrity and quality—the ingredients that were saving the company.

He then thanked them for the car and said, 'But I have such confidence in you now that I will take the next car off the line.' He had shown that he would place his trust in them, but also that they should not mess with him.

Story 2

One night, in the US Midwest, a communication tower blew over in a storm. The tower was vital in keeping a certain parcel service operating, especially because most of the service's shipments were done overnight. A junior employee (some versions of the story suggest it was the night watchman) recognised the importance of the tower and also noted that everyone was empowered to do whatever was necessary to deliver parcels overnight. So he hired a helicopter to reach and reinstall the tower, and by morning the tower was back in operation.

Story 3

The senior management of a company set about instituting a change programme. They reckoned that change would only start when a significant proportion of the employees saw what the next step might be as well as the benefit of making change.

So they established a dialogue (which in itself was a cultural shift from a top-down one-way mode of managing) about what was wrong with the company and what it had to do to beat the competition. They legitimised dissent in the name of finding better ways forward, based on an honest appraisal of the state of the business.

They then set up natural work teams around key problem areas (i.e., teams of people who had a stake in solving each problem), gave them a budget and authority to ask any question and explore any solution, and set a time-frame for them to report back with practical solutions. The process yielded not only solutions to problems but also different ways of doing things, and this established a more team-based, problem-solving, and participative culture.

A variation to this approach is to follow up stories with a wishful thinking exercise. Ask yourself as an organisation, 'What are the stories that I would want a senior manager or an employee to tell about us when they retire?'

5. Visual Techniques

There are a range of techniques that are related to the concept of linking language with visual images. Using visual images can stimulate a different part of the brain.

a. Mind Maps

Mind-mapping is a technique that has been around for centuries in various guises. Leonardo da Vinci's notebooks were full of text and pictures combined in ways that we would recognise now as mind maps.

More recently, Tony Buzan (10) developed and promoted the mind map as a tool for note-taking and planning. Because it encourages the organisation of information in a non-linear way, it mirrors and stimulates our brain's associative thinking capacities. As a result, we can find ourselves making a number of unexpected associations between topics that can lead to creative ideas.

Mind-mapping is a flexible process that allows for ongoing addition without the need for re-drafting.

The steps to creating a mind map are as follows:

- Take a blank piece of paper (and turn it on its side so that it's landscape view)
- Write the central theme, idea, or issue in the centre of the page
- Use key words or phrases to develop the main point on lines or 'limbs' or 'branches' radiating from the central point
- Add further ideas or details of elaboration/clarification as further branches off the main limbs. You can build on these further and further, and come back to them as more ideas strike
- Using different colours and symbols/pictures can help ensure clarity on the mind map, as well as stimulating the brain

Below is a simple example. Of course, there are also many online programs that will help you build a mind map.

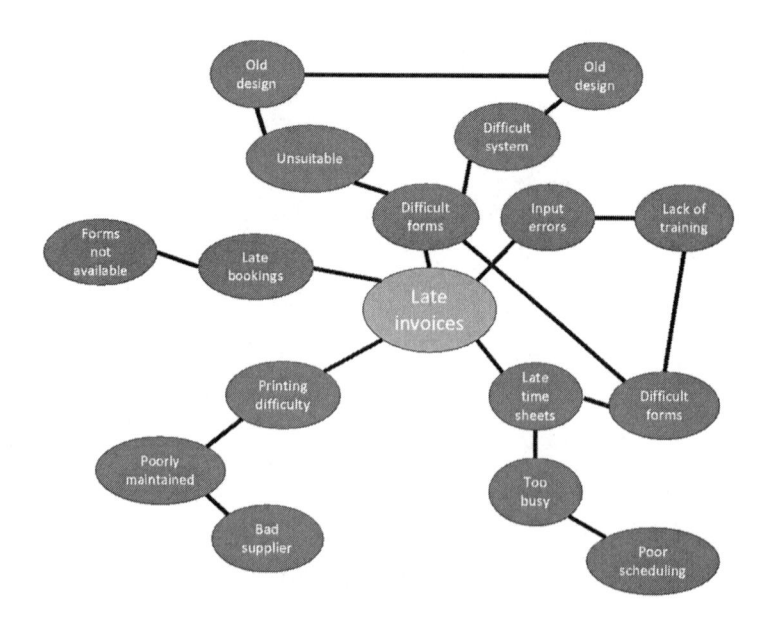

What are some of the advantages of a mind map?

- Simple to use.
- The central idea is clearly defined.
- The relative importance of each theme can be seen; fundamental themes are closer to the centre, and elaborations of these themes are further away from the centre.
- Links between key concepts are recognisable because of physical proximity and connection.
- The associative nature of the map, mirroring the associative nature of our brains, encourages new connections that might not have arisen from a linear approach.
- New information can be added without the need to re-draft the mind map.
- The radial nature of the map allows for associations in any direction.
- A lot of information can be stored and conveyed on one map.
- Recall and review is quicker and more effective than a linear list.
- It can be a fun, engaging and energising process, which is particularly useful for group work.
- The use of different colours for limbs and branches keeps points clearly separate.
- Adding images and pictures helps the map come to life and stimulates the visual aspects of our brains.
- The best mind maps are the ones you do yourself according to your own structure and process.

b. Fishbone Diagram

A Fishbone diagram is an example of a *cognitive map* (as is a mind map), which is a way of graphically displaying the connections between a collection of ideas, concepts and words. Fishbone analysis (also known as 'cause-and-effect analysis' or 'Ishikawa analysis' after management expert Kaoru Ishikawa who invented the technique) is a structured way of getting to the heart of the real causes of a problem or situation. It provides a method of breaking a problem down into more manageable elements (11).

To produce a Fishbone diagram, take the following steps:

- Identify and agree upon the problem statement and write this in a box at the edge of the page. This is the head of the fish.
- Draw a backbone from the head of the fish along the page.
- Identify 4-6 areas that are likely to have a bearing on the problem.
  - The areas used most frequently are: people, procedures, materials, and equipment
  - Other areas that might be used include: policies, processes, machinery
  - Don't worry too much about choosing the 'right' categories, they're only to help organise the suggestions for issues and their causes
  - Write each category on one or other side of the fish and connect them to the backbone (see example below)
- Spend 5 minutes (group members should do this individually) writing down all the causes of the problem that come to mind, bearing in mind the key categories identified, and also referring to the process map[12] (the sequence of steps in a process) if one has been created.
- Focus on each of these ideas, one at a time. Group members can take it in turns to offer one of their ideas.
- Each suggestion is put on the chart connected to the main 'bone' for the relevant category.
- For each suggestion keep asking 'why' until all the root causes of this suggestion have been unearthed and logged on the fish diagram. Each suggestion then has a line with an arrow pointing from one cause to the next.
- Then move on to the next suggestion until all suggestions have been explored.
- Finally, examine each root cause, and brainstorm potential solutions.

The '5 whys' technique (explained above in the section on reframing), where the facilitator asks 'why' until the root cause or issue is uncovered, can be particularly useful when conducting a fishbone exercise, and is often used.

---

[12] A process map is a set of steps that charts the sequence of events in a specific process. The map identifies the interdependencies of events, the choices that have to be made and the final outcome.

Fishbone diagram

[Diagram showing a fishbone/Ishikawa diagram with major and minor causes branching off a central arrow pointing to "Problem"]

c. Picture Associations

It is said that a picture is worth a thousand words. It is likely that pictures are processed in a different part of the brain to written words, or verbal sounds, so using a picture as a stimulus for brainstorming may well bring different results.

Group exercise:

- The facilitator presents a picture that is related to the issue at hand.
- Participants say whatever words spring to mind, and use this as a springboard to discuss elements of the issue, or brainstorm solutions to the issue.

Variations include doing the following:

- Present a range of images instead of just one.
- Include abstract, beautiful, shocking images, or images that appear at first glance to be unrelated to the issue.
- Present an 'intriguing' or ambiguous picture and ask people to describe what they think is going on, or what happens next, or what the consequences might be. Then use these answers as the springboard for a brainstorm of solutions.

- Present a sequence of pictures; e.g., a crash sequence, and ask people to explain what was happening, and why one picture led to the next. This encourages close observation of the details.
- Use a picture of an animal and explore how they might approach a human task (and perhaps do it better). Use the comparison with a human approach to generate new ideas.
- Use specific picture-related props. For example, Roger von Oech's Creative Whack Pack includes cards showing pictures and stories that can be passed around a group to stimulate thinking.

d. Drawing the Problem

This technique can be useful in a group setting. Everyone individually draws the problem, or particular aspects of the problem. Then the group discusses the various drawings and aspects of the problem that are depicted. This can highlight many different aspects of the problem, as it reflects the different perspectives of each individual. This might then provoke new insights within the group as to how to solve the problem. It could also lead to the creation of a combined group-view picture.

A variation is to ask the group (or different groups) to interpret each drawing with no verbal input from the creator of the picture. This can lead again to new insights and associations.

e. Storyboarding

This technique arose from cartoons, and the advent of movie-making. The cartoon strip or movie storyboard—a series of chronological pictures that tell the story—can be used to tell all manner of stories, not just films, cartoons or novels, but also business proposals or PhD theses.

The technique can be used to draw the problem and then potential solutions to the problem. For example, a company in a service industry might use this approach to improve its service provision with the following approach:

- Imagine that the company has just won a prestigious award 5 years from now for the most innovative service provision.
- Then draw out what the aspects of this service provision look like and the steps in the story to get there.

Exercise:

For a fun creativity exercise, draw a number of boxes across a page like this:

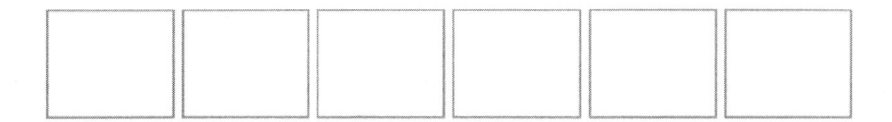

Imagine you are coming up with themed episodes for a cartoon about a warring cat and dog. In each box, draw a potential scenario. One approach to this exercise is to place time pressure on the process—e.g., fill in as many boxes as possible in 5 minutes.

This encourages the participant to attempt to generate a large number of ideas for the boxes, without worrying too much about whether the idea (or the drawing) is actually any good or not. As we discussed in the section about brainstorming, generating a large quantity of ideas without initially worrying about the quality of these ideas is an important part of the idea generation process.

f. Shield Exercise

This exercise is a useful technique to help a team find its identity and articulate what it is about—its past, its present and its potential future. Groups of all kinds—countries, families, sports teams—have represented themselves pictorially over the centuries.

The team identifies an animal and a motto that best represents them, and considers how to complete the four quadrants of the shield that represent the team's past, their present, their hopes for the future, and their fears for the future (see the shield below).

This exercise is designed to help teams gain a shared understanding of their identity and their purpose, and liberate their thinking by putting on the table various aspects of the team's identity that often go unsaid. This exercise can be particularly useful for teams who are grappling with transition and change.

Exercise:

- Split the team into groups of 4 or 5.
- Each group draws the outline of a shield (see example below).

- The group then works together to fill in the boxes on the shield with pictures and text that best represent how they feel.
- After 15 minutes, ask the groups to describe their shield to the other groups, and to compare and contrast them.
- It can be useful to keep the shields visible, and possibly to continue to add features to the shields as new insights come to light.
- Groups who are experiencing transition can also be asked to draw the journey they face through this transition on the four quadrants e.g., where are we starting from, where do we want to get to, what good things will we find there, and what obstacles will we have to overcome along the way?

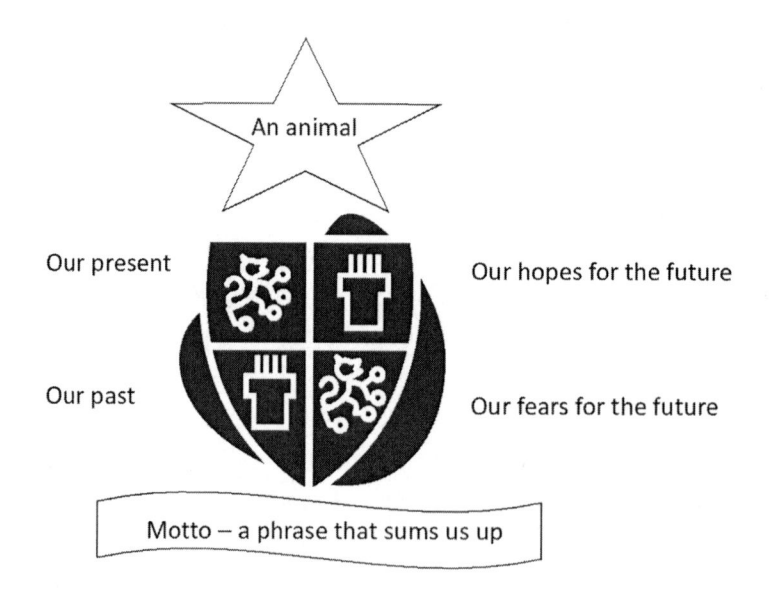

An animal

Our present

Our hopes for the future

Our past

Our fears for the future

Motto – a phrase that sums us up

g. Visual Success Map

A visual success map aims to provide a very clear overview of what success looks like and what are the necessary steps to achieve it. There are six steps to plotting a visual success map.[13]

---

[13] The Visual Success Map has been developed by Farren Drury of Go Make It Yours (12), and software to facilitate the creation of visual success maps is available at www.gomakeityours.com

Step 1: What does success look like?

- Write a headline for what you will have achieved at the end of the project.
- What is life like at the end of the project? E.g., at work, at home, on a personal level.
- What is different for your team and for others around you?

Step 2: Describe the destination in terms of what you will do to accomplish it.

- E.g., the human resources project group will gather data and analyse the results, propose actions, and implement these actions to achieve a greater than 20% reduction in labour turnover.

Step 3: Identify the major themes or lines of work that must be pursued in order to achieve success.

- E.g., for the human resources project, these themes might include:
  - Data analysis
  - Policy development
  - Organisational development
  - Recruitment
  - Training

Step 4: Develop activities that are stepping stones to achieving the outcomes for each line of work.

- E.g., Data analysis might include the following:
  - Establish data items required
  - Identify location and accessibility
  - Check data integrity
  - Determine analyses and modes of presentation
  - Produce results
  - Iterate with further analyses

Step 5: Analyse and chart the detailed tasks required to achieve each stepping stone.

Step 6: Decide on the best course of action…and execute the plan.

- Win support for the plan, review progress towards it, and keep the plan alive
- Monitor and celebrate your success

Visual Success Map (VSM)

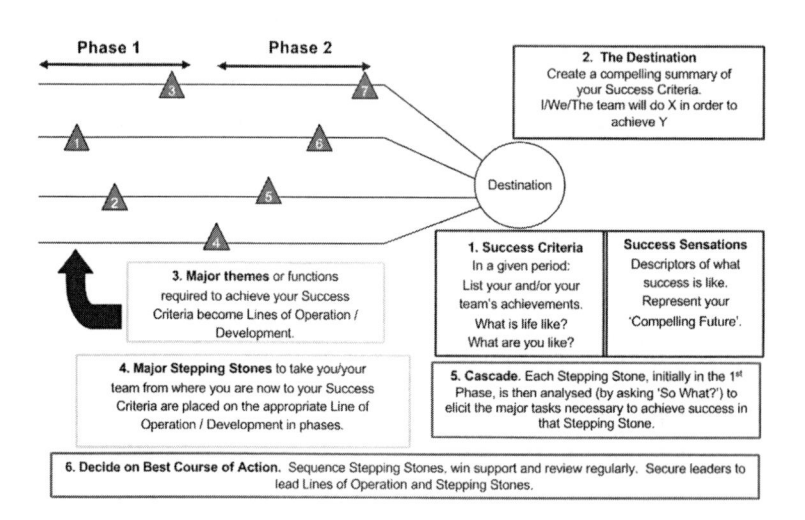

Phase 1 / Phase 2

2. The Destination — Create a compelling summary of your Success Criteria. I/We/The team will do X in order to achieve Y.

Destination

1. Success Criteria — In a given period: List your and/or your team's achievements. What is life like? What are you like?

Success Sensations — Descriptors of what success is like. Represent your 'Compelling Future'.

3. Major themes or functions required to achieve your Success Criteria become Lines of Operation / Development.

4. Major Stepping Stones to take you/your team from where you are now to your Success Criteria are placed on the appropriate Line of Operation / Development in phases.

5. Cascade. Each Stepping Stone, initially in the 1st Phase, is then analysed (by asking 'So What?') to elicit the major tasks necessary to achieve success in that Stepping Stone.

6. Decide on Best Course of Action. Sequence Stepping Stones, win support and review regularly. Secure leaders to lead Lines of Operation and Stepping Stones.

6. Attribute Combinations

a. Morphology Strips

Morphology strips are designed to challenge convention, habitual wisdom and automatic assumptions. They aim to highlight every potential combination of key ingredients in the situation, rather than just the most usual or expected combinations.

The process is based on the children's game of mixing clothes from different characters, and also on the game of 'consequences'.

As an example: Imagine you had to write an episode for a television programme.

- Take six strips of card
- In the following steps, write the same number of words (aim for 5 or 6) on each strip
- Write the names of the key characters in the programme along the first strip
- Write action words along the second strip
- Write physical objects along the third strip
- Write places and locations along the fourth strip
- Write characterisation along the fifth strip
- Write feelings or emotions along the sixth strip

The words that you have written on the six strips are the 'ingredients' for your episode. Place the strips next to each other so that the words are in columns, one above the other. Slide the strips back and forth, observing the different combinations of 'ingredients' that arise in each column as you do so.

For a particular product, you might approach the exercise in the following way:

Characters = users of the product
Action = how they might use it
Object = what might they use the product in conjunction with?
Place/location = where might they use it?
Characterisation = what are the most important features or functions?
Feelings/emotions = how would they feel using the product?

b. Ideas Matrix

If there are only two key attributes, each with a number of different possibilities, you can explore every potential combination by using a grid—with a list of options for each attribute along each axis.

For example, a trainer might start with a list of activities and participants.

Look at each point of the grid, representing each potential combination of attributes, and see whether it yields any interesting or creative possibilities, or possibilities that have not been tried before.

As another example, a producer who is exploring making new TV programmes for youth audiences, might identify two key attributes, 'genre' and 'audience segment'. She then lists the items within these categories (e.g., 'game shows' as a possible 'genre') in a matrix and then explores all the potential combinations that arise. There are combinations that are already in use (marked ✔ in the table) but the exercise highlights untapped combinations that might be worth targeting with new programming.

	Under 5 years	5-10 years	Tweens	Early teens	Late teens
Movies	✔	✔	✔	✔	✔
Game shows	?	?	✔	✔	✔
Documentaries	?	?	?	✔	✔
Talk shows	?	?	?	?	?
Reality shows	?	?	?	✔	✔
Etc.					

7. Serendipity

An Arabian story recounts the journey of the Prince of Serendip, who wanted to discover wisdom. His wise men told him to journey through many lands and that he would find wisdom by doing so. He travelled far and wide, but upon his return he berated his officials, telling them that he had not found wisdom.

However, the wise men asked him to recount the details of his journey and describe all the things he had seen and heard along the way. As the prince told the story of his journey, he realised that he had learned many things through his travelling and had returned much wiser.

This story has given the English language the word 'serendipity' as a label for when things occur tangentially or by chance. In the prince's case, the wisdom he craved was tangential to the travel adventures he experienced.

Innovation often occurs in tangential ways. For example, mistakes in research or experimentation can lead to unexpected innovation, and there are many instances of inventions that have impacted our lives in significant ways, but were actually mistakes or accidents.

Examples include the microwave, which was discovered during research into radar engineering when a chocolate bar melted in the engineer's pocket; Viagra, which was originally developed as heart medication; chocolate chip cookies, which were the result of an attempt to make regular chocolate cookies using pieces of a regular chocolate bar instead of baker's chocolate—the chocolate pieces failed to melt in the cookie dough, as expected, and the chocolate chip phenomenon was accidentally born.

Likewise, war-time innovation for military purposes can lead to numerous creative applications in peace-time. Thermoplastics—such as Teflon or nylon—were a direct result of the innovation required to create the heat shields necessary for space missions. The system of triage, used to prioritise medical cases, results from experiences on the battlefield.

As well as simply being open-minded and alert to the opportunities that mistakes can present, there are a variety of ways that serendipity can be harnessed to lead to creative outcomes.

a. Improvisation

This is most evident in music and other forms of performance. Jazz music frequently contains sections of improvised soloing, where the soloists play the music that comes to them in the moment, rather than playing pre-planned music. Each time the musician plays an improvised solo it will be different.

In the case of live performance, these improvised solos are fleeting—heard by those present then lost forever. But some artists have explicitly harnessed improvisation in more concrete settings. Miles Davis was known for not practicing pieces with his recording ensembles so that any solos played during the recording would be truly improvised (rather than re-hashed versions of solos played during the practices), and would retain their creative vibrancy as a result.

The pianist Keith Jarrett recorded a series of concerts in which he played solo piano and improvised the entire performance. His Köln concert is widely regarded as a masterpiece, and yet apparently it nearly didn't happen, as the concert grand piano he had requested was not provided, with only a baby grand available in its place.

Furthermore, the piano was not in particularly good shape, and some of the memorable elements of Jarrett's performance may have resulted from this misfortune and the constraints it brought. He later reflected, "What happened with this piano was that I was forced to play in what was—at the time—a new way. Somehow I felt I had to bring out whatever qualities this instrument had."

Improvisation also takes place when people experiment with using objects or processes in ways other than their intended uses, frequently in times of necessity. Any student of the film Apollo 13 will readily recognise what can be done with scraps of material when survival is at stake.

How can improvisation be explicitly harnessed? In the case of a performance, it may be by rejecting pre-planning, and going with what springs to mind in the moment. This is not to say that there is no pre-planning whatsoever, however. Much successful improvisation comes about because the improviser has a vast amount of knowledge and experience in the particular field. As a result, he or she has a wide range of tools and raw materials with which to improvise.

b. Randomness

Randomness is another way of introducing chance elements to a situation. As with improvisation, there are various examples in the creative arts of randomness being explicitly harnessed for creative output.

In 1979, filmmaker Andy Voda used the process of flipping a coin or rolling a dice to dictate many of the decisions in the movie *Chance Chants*.

The renowned avant-garde musician John Cage has composed music by superimposing star maps on blank scores, by rolling dice to dictate notes used, and by preparing open-ended scores that depend on spontaneous decisions made by the performers (a form of improvisation).

Possibly his most famous work is 4'33" (4 minutes, 33 seconds), which comprises exactly 4 minutes and 33 seconds of 'rests' (i.e., the musicians remain silent). However, the piece is not designed to be 4 minutes and 33 seconds of silence, so much as 4 minutes and 33 seconds of the random ambient sounds that might occur during the performance of the piece—a whir of an air conditioning fan, a cough from someone in the audience, the turning of a page of sheet music.

In this way, each performance is different, and is entirely random, based on the ambient sounds in the moment of performance.

Random chance events can also be the catalyst for the flash of insight that comes during the illumination stage (after all the hard work of preparation, experimentation and incubation has taken place). It is often this random chance event and the resulting insight that constitute the creation story for a particular idea.

In 1862, the German chemist Friedrich August Kekulé was relaxing in front of a log fire. He had been working on the chemistry of carbon-based compounds and the theories behind these compounds for many years. As he dozed in front of the fire, he saw in the flames a vision of a snake catching its own tail. This random occurrence inspired his insight that the structure of benzene contained a ring of carbon atoms.

As we discussed in Chapter 1, these insights can often occur during or after a period of relaxation or distraction. It is therefore worth taking a break from creative work periodically to do something different—read a book, watch a movie, go for a walk—while the subconscious mind continues to work on the problem. This can lead to insight arising when you least expect it.

c. The Unconscious and Dreams

Another way that people have tried to harness the nature of randomness or serendipity is by examining their dreams for creative possibilities. This is easier said than done, but might throw up some novel scenarios, or combinations of elements, that the conscious mind might not yet have happened upon. The challenge can be remembering the dream, however!

As we've already remarked, Paul McCartney famously woke up with the melody to the song *Yesterday* in his head and used the lyric 'scrambled eggs' to accompany it until he could write something more meaningful.

The artist Salvador Dali had an interesting technique to harness the power of his dream-state. Legend has it that he would settle down for a nap holding the end of a spoon lightly in his fingers as the spoon balanced on the side of a glass.

As Dali drifted off to sleep, so his grip on the spoon would loosen, causing it to fall into the glass with a clatter, thus waking Dali. He would then try to recall the contents of his mind at that point on the precipice of sleep and use it as inspiration in his art.

8. Research

As we have previously discussed, knowledge or expertise in a subject matter is an important part of the creative process. Sometimes, the methodical acquisition of further knowledge in a particular area through research, and a critical eye turned on this research, can yield new possibilities.

Alexander Fleming had seen how many soldiers had died in the First World War from sepsis and for the next 8 years he looked for a cure. By 1928, he was already well known from his earlier work, and had developed a reputation as a brilliant researcher.

On returning from a holiday, he noticed that one of the cultures he had left on his lab bench was contaminated and he identified the mould as from the genus Penicillium, from which he developed penicillin.

One sometimes finds, what one is not looking for. When I woke up just after dawn on September 28, 1928, I certainly didn't plan to revolutionise all medicine by discovering the world's first antibiotic, or bacteria killer. But I suppose that was exactly what I did. – Alexander Fleming (13)

9. Combined Approaches

As the field of creativity has gathered increased attention and research, various theorists have put forward approaches that combine a number of the different strategies that we have already discussed. In codifying these combined approaches, the theorists seek to construct effective methods to generate a quantity of ideas, as well as fresh perspectives and insight, regardless of the issue at hand.

a. TRIZ

Genrich Altshuller's (14) pioneering work in the former Soviet Union examined some 200,000 patents and identified a number of principles which underlie technical innovation. They are collectively called TRIZ (an acronym for a Russian phrase describing tasks that relate to invention). Here are a few of the 40 approaches he proposed:

- Segmentation—divide an object into independent parts
- Extraction—remove only the problem part of a system
- Local quality—get each part to do its job optimally
- Asymmetry—create a different balance between the parts
- Consolidation—'consolidate in space homogeneous objects or objects destined for contiguous operations'
- Universality—look for other ways and contexts to use the same object

There is considerable detail involved in TRIZ and it requires specific study before application. It is most useful when looking at physical problems, such as those found in manufacturing and engineering. A typical approach would be:

1. Bring together a problem-solving team
2. Use a problem-solving process (see Chapter 7)
3. Study the 40 TRIZ approaches
4. With reference to TRIZ techniques, explore the different options for solutions that might emerge

5. Experiment with solutions to see what works practically

b. Roger Von Oech's Creative Solutions

Roger von Oech's (15) work, incorporating some of the teachings of the Greek philosopher Heraclitus, also identified some 40 creative strategies than can be deployed in any situation.

A selection of the approaches for gaining greater insight include:

1. Look for patterns in phenomena and the meanings that patterns can convey.
2. Don't dismiss something that you don't understand or that seems wrong.
3. Be wary of your past successes—they make you less open to ideas that challenge the route to that success.
4. Anticipate criticism and find ways of working with it constructively.
5. Stability, whilst often desired, works against innovation and change. Find ways of creating turbulence.
6. Stay with ambiguity. Early decisions increase comfort but may result in choosing the easy, simple solutions and avoiding the longer lasting ones. Always look for multiple solutions.
7. Go with the flow of events. Stopping resisting events may offer fresh opportunities.
8. New problems arise when situations become large. Keep things small where possible. They are more easily controlled and more manageable.
9. Things sometimes have to become worse before they can get better (the doctor may need to inflict pain to bring about ultimate healing).
10. Do the opposite of what you are currently doing. (Speak quietly to gain attention, admit helplessness to gain influence.)
11. Challenge conventional wisdom. What assumptions do we make that might be wrong?
12. Look for the missing elements, 'the pauses between the musical notes', 'the things we don't talk about'.
13. Don't just solve problems, go beyond and seek the opportunities.
14. Reframe the situation—use different language to redefine a situation. Find a different position to view the problem.
15. Look for opportunities in 'mistakes'.

16. Forgive the past—don't hold on to past grievances as the reason for current action.
17. Lighten up! See the situation as a game and an opportunity to have fun. You might be surprised by what emerges when freed of the constraints of having to be serious.
18. Understand what deceptions and self-fulfilling prophecies are operating in your life—for good or ill.
19. Stand back from the problem. The closer you get to it, the harder it can be to come up with the novel solution.
20. Counter the dominant force to allow other voices to come through. 'The stars are only apparent when the sun sets'.
21. Foster curiosity about the world around you. Take ideas from fields of enquiry that are not your own.
22. Speculate about the hypothetical. What if…
23. Use constraints—time, money, etc.—and the threat of real or imagined competition to force more ingenious solutions.

c. TILMAG

German creativity consultant Helmut Schlicksupp created the TILMAG method, which provides a structured approach using association to provide insights and solutions.

TILMAG is a German acronym, referring to the process of deriving the elements of ideal solutions through forming associations and discerning commonalities.

The process is more structured than other tools and is similar to attribute combination techniques.

The steps are:

- Define the problem to be solved and then use brainstorming to find all possible solutions.
- Define the elements or attributes that the ideal solution must possess:
    o Specific to the problem
    o Described in a positive way
    o Brief but precise
    o Related to customer demands (16)

- Construct a matrix with these elements on each side of the square and seek to identify any solutions or associations that come to mind from the combinations of each element with another (see the Ideas Matrix earlier for a similar approach). These ideas can be as wild and wacky as you like.
- Identify the underlying principles behind each solution or association and reconnect them back to the problem.
- Prioritise the solutions that have arisen.

d. SCAMM/ESEAM/SCAMPER

SCAMM, ESEAM and SCAMPER are all acronyms for a series of steps designed to prompt fresh perspectives through approaching the problem in a number of different ways. The different approaches therefore combine the steps below in the following ways:

SCAMM = Substitute, Combine, Adapt, Magnify, Minimise
SCAMPER = Substitute, Combine, Adapt, Modify/distort, Put to other purposes, Eliminate, Rearrange/reverse
ESEAM= Eliminate, Simplify, Export, Automate, Measure

The individual steps:

SUBSTITUTE: Consider substituting one element of the process for another—or for something else entirely.

- E.g., materials, personnel, location, mode of thinking (positive vs negative)

COMBINE: What parts of the situation could be combined?

- Where can combinations lead to synergy?
- E.g., can we combine customer research with employee attitudes to get a better picture of where an organisation's problems lie?

ADAPT: What element(s) of the situation/issue/product can we adapt to make a positive change? How can we adapt what already exists?

- E.g., how can we change existing policies, products, materials, technology, people, etc.?

MAGNIFY: What can we magnify to create solutions?

- E.g., how can we build on success, or build momentum by taking successful small ideas and promoting them, or rolling them out more broadly? How can we find small pockets of success and good practice, and spread them to less successful areas?

MINIMISE: What can we reduce to bring about change and insight?

- E.g., How can we reduce problematic processes?

MODIFY/DISTORT: How can we change part or all of the current situation, or distort it in an unusual way?

- By forcing ourselves to come up with new ways of working, we are often prompted into an alternative product or process.

PUT TO OTHER PURPOSES: What elements of the situation can we use in other ways?

- What can we re-use from elsewhere to solve our problem?
- What other markets can we approach with our product?

ELIMINATE: What can we eliminate to bring about success?

- What might happen if we eliminated various parts of the product/ process/ issue, and what would we then do in that situation?
- E.g., Eliminate wasted time, effort, materials, expenditure, etc.

REARRANGE/REVERSE: What elements of the situation can we rearrange or reverse to bring about solutions?

- E.g., can we reverse any aspects of the manufacturing process?

SIMPLIFY: How can we simplify the problem or the process?

EXPORT: What can we export to other people or locations in order to increase productivity or efficiency?

AUTOMATE: What elements of the process can be automated?

MEASURE: What are the critical processes that we should focus on, and what are the measures of improvement or success?

10. Observational Approaches
a. Customer Ideas

Organisations can learn a lot from their customers, but frequently fail to engage them in the right way to do so.

Organisational theorist Tom Peters studied organisations that delivered superior customer service and found that "they learn from the people they serve. They provide unparalleled quality, service and reliability. Many of the most innovative [organisations] got their best service ideas from customers. That comes from listening closely and regularly to what they say" (17).

How to elicit accurate information from customers is another challenge. Sometimes, a simple questionnaire isn't enough. Coffee chain Starbucks launched a customer-oriented website to encourage customers to submit suggestions, improvements, requests—from customer service to new products—an approach that has been used by a number of other companies.

Unilever has an 'Open Innovation' portal where they invite anyone to submit ideas for certain areas of the business. And various bands, including Radiohead, have made elements of their music available to their fans and encouraged them to remix or reconstruct their songs.

b. Empathic Design

Empathic design entails observing the customer as they use the product in order to investigate the different approaches they have to product-use, as well as contextual factors that might promote (or hinder) product-use.

When a shampoo company spotted that people sometimes washed their hair with shampoo twice over for a really clean wash, they inserted instructions on the packaging to encourage everyone to do so ("rinse and repeat"). As a result, customers began using double the quantity of shampoo, and sales rose significantly.

Tool company Black and Decker used empathic design to develop the paint-stripper tool. One day, the wife of a manager at the company left her hair dryer turned on full blast and pointing at some paintwork that bubbled under the heat. The manager saw this and realised that the hairdryer would actually make a good heat gun for paint stripping.

An engineer at the company then recognised that the hairdryer was actually simply composed of a motor/heating element and a fan—common elements in many products—and with minimum re-engineering to the common elements of an existing product (the heating element was made a bit stronger), a new and successful product was born.

Other approaches that companies have taken include observing the modifications that customers have made to products. Representatives from Harley Davidson motorcycles have attended Harley Davidson owners' gatherings and conventions to observe the customisations that owners had made to their bikes as a source of ideas for product enhancements and accessories.

11. Selection of Ideas

Once ideas have been generated through the use of idea generation techniques, the next step is to decide which ideas to move forward with.

To do so, relevant criteria must be set for evaluating the ideas. There might be a single criterion, or a number of criteria. Sometimes, a selection grid can help. In this instance, it can be helpful to use criteria that elicits Yes/No, True/False, High/Low answers.

a. Selection Grid

Example: An accounts payable team evaluated its performance and conducted a brainstorm to elicit ideas for how best to improve its performance. They chose two clear stand-out ideas, but needed to narrow these two down to one idea. They used a selection grid to evaluate each idea, as well as to identify any further action that needed to be taken in order to successfully implement the idea.

Improvement opportunities	Do we have control over key variables?	Is it doable in our allotted time?	Are the right people on the team?
1. Eliminate missing paperwork	No	No	No
2. Ensure correct vendor numbers on invoices	Yes	Yes	No

As a result of this evaluation, the team made arrangements to bring the critical people onto the team, and then proceeded with option two.

b. Multi-Voting

When it is necessary to reflect each participant's view of the importance of each item on a list of ideas, multi-voting can be helpful.

The steps are:

- Eliminate duplicates and items that are universally agreed by the group as unimportant.
- Summarise the selection of items to be considered for further work.
- Each group member writes down a rank order, 1 being the least important; e.g., if there are five items, a member might rank them as A - 4, B - 5, C - 3, D - 1, E - 2.
- The rankings of all team members are combined by adding up the scores for each item, e.g., for item A, there might be ranks 4, 3, 2, 5, etc. totalling 14. For item E, there might be 2, 4, 3, 1 totalling 10. So item A would be more important overall and the group would work on that before moving on to others as time allowed.

An alternative approach is weighted multi-voting. Instead of ranking items, a score of say 100 points is allocated across the items by each participant, e.g., a member may allocate points as follows: A - 40, B - 10, C - 20, D - 5, E - 25 (total = 100). Then the rankings are combined as before.

Remember that the process is about getting agreement on the choice of a particular item to consider further. Too much focus on the method might deflect participants from the actual work of choosing an appropriate idea to pursue.

c. Impact Analysis

- Each group member describes how the selected problem impacts their work.
- The group confirms that the problem is relevant to the group members (if it transpires that the group feels that the problem isn't all that relevant, they can decide not to proceed and turn their attention to other problems).
- The group uses the grid below to prioritise ideas for action, evaluated for whether they are high or low impact and difficult or easy to implement. Those that are high impact and easy to implement—the master strokes—are for prime action.

  Whilst quick wins (low impact, easy to implement) may not be so impactful, they are easy to implement and can therefore be used quickly to build morale and momentum, as well as a belief that change is possible.

  Those ideas that are high impact but are difficult to implement—the slow burn—may be able to give an organisation a competitive example (as they are difficult for others to follow), and so should be pursued over time. As for low impact ideas that are difficult to implement—ignore them!

Impact analysis

	Low ease of implementation	High ease of implementation
High impact	Slow burn Competitive edge	Master strokes
Low impact	Dodos	Quick wins

12. Implementation

Once ideas have been generated through the use of idea generation techniques, and narrowed down to a select few, the next step is to consider implementation. This is where the process moves from creativity to innovation.

a. 'The Pebble in the Shoe'

Change rarely comes about without some incentive. The expression, "Necessity is the mother of all invention", sums this up nicely! Other expressions include the 'burning platform' that you need to get off as quickly as possible; the 'pebble in the shoe', an annoying presence that won't go away until you make a change.

It is possible to elicit a situation within an organisation that acts as a 'pebble in the shoe' in motivating change. Some examples are as follows (use the 'applicability to you' column to assess your own situation or organisation):

Intervention	Applicability to you
Identify and amplify an external threat	
Create a crisis	
Model self-criticism and encourage others to do the same	
Use customers to push for higher standards	
Survey internally to find sources of dissatisfaction or problem areas	
Use external benchmarks with better organisations to highlight areas for improvement	
Provide forums for open discussion about improvement	
Tell stories about how the best organisations and people are continually looking for ways to improve	
Recruit and protect mavericks and those who do not fit the conventional mould	
Move people into new positions who will look afresh at the old problems	
Recruit new people who bring with them no 'baggage'	
Use external agencies to identify areas for change	
Send key opinion formers on external study tours	

b. Force Field Analysis[14]

It can also be very useful to conduct a force field analysis to identify where the areas of resistance to change might lie, as well as those forces and factors that will promote and sustain the change.

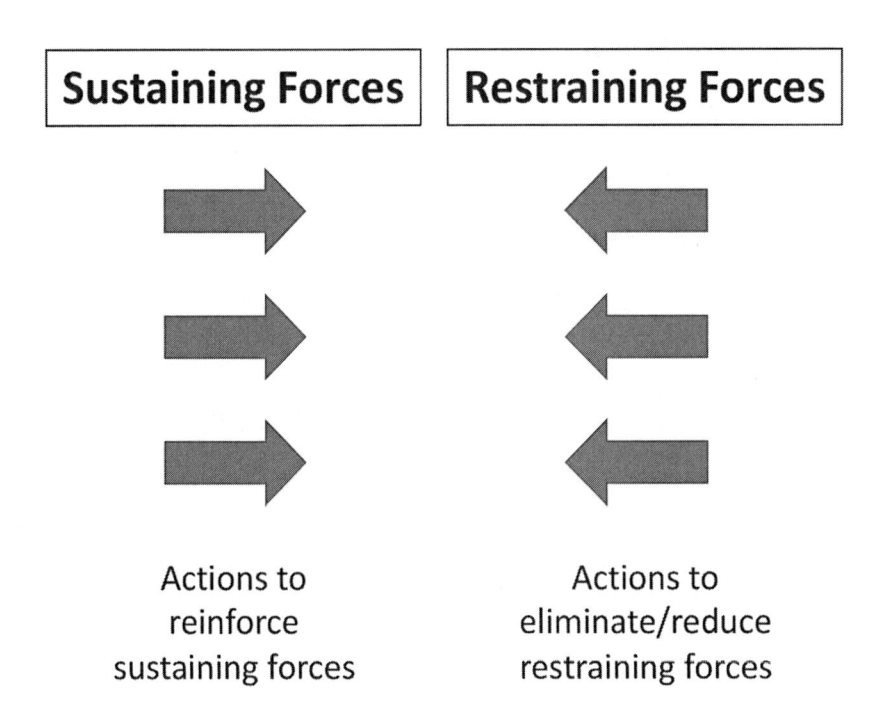

Sustaining Forces	Restraining Forces
➡ ➡ ➡	⬅ ⬅ ⬅
Actions to reinforce sustaining forces	Actions to eliminate/reduce restraining forces

---

[14] Force Field Analysis was developed by psychologist Kurt Lewin.

Chapter 6
Creating an Organisational Culture for Creativity and Innovation

Introduction

'It's not in the rules.' 'You need permission for that.' 'We don't do it like that.' These rebukes are the stuff of organisational life, mantras that are heard in the corridors, offices and boardrooms of businesses every day, and used to shut down ideas, suggestions, changes, innovations and just about anything that might challenge the status quo.

These phrases have often emerged as a result of some previous business failure, and are now tools in the process of applying stricter controls for the future. Our individual and collective fear of failure (and its costs), and our comfort with the status quo, reinforce the power of these controls.

However, when we are considering the organisational conditions that might be most effective for creativity and innovation to emerge and flourish, it's worth being mindful of another old adage, often attributed to the infamous British soldier Harry Day: 'Rules are tools, a guide for the wise, the blind obedience of fools.'

As we mentioned in the introduction to this book, IBM conducted a major survey of more than 1500 Chief Executive Officers from 60 countries across 33 industries to assess their view of the skills and capabilities required for future success in navigating an increasingly complex world (1).

The survey found that these CEOs placed creativity as the number one capability, above rigour, management discipline, integrity (ha!) or even vision.

Less than half of the CEOs surveyed believe their enterprises are adequately prepared to handle a highly volatile, increasingly complex business environment. CEOs are confronted with massive shifts—new government regulations, changes in global economic power centres, accelerated industry transformation, growing volumes of data, rapidly evolving customer preferences—that, according to the study, can only be overcome by instilling creativity throughout an organisation.

More than 60 percent of CEOs believe industry transformation is the top factor contributing to uncertainty, and the finding indicates a need to discover innovative ways of managing an organisation's structure, finances, people and strategy.

As a result, creative leaders:

- Expect to make more business model changes to realise their strategies
- Invite disruptive innovation, encourage others to drop outdated approaches and take balanced risks
- Consider previously unheard-of ways to drastically change the enterprise for the better, setting the stage for innovation that helps them engage more effectively
- Are comfortable with ambiguity and experiment to create new business models
- Score much higher on innovation as a crucial capacity and more of them expect to change their business models
- Are courageous and visionary enough to make decisions that alter the status quo
- Will invent new business models based on entirely different assumptions

Evidently, they have to throw out the old rules, give 'permission for that' (whatever that is) and give people freedom to 'do it like that'. So how should business leaders approach this organisational challenge?

The Challenge of Organisation Design

Organisation design, and the resulting organisational culture, have been around as long as humans have been organising themselves collectively to achieve their various aims. The Bible records an early example of organisation design when the prophet Moses is advised by his father-in-law Jethro to divide the work and delegate to "capable men from all the people… That will make the load lighter because they will share it with you" (2).

This simple instruction illustrates an underlying desire of groups—to establish hierarchy and specialisation, to push work down the hierarchy, and to divide work into prescribed roles, based on an assumption that routine work is more efficiently managed if divided up.

Given the routine and systematic nature of much work, and its dependence on hierarchy for direction, order and control, this approach has worked successfully for a very long time. In the 19th century, the approach really came into its own as the mainstream paradigm for organisational life, as the concepts of 'scientific management' and bureaucracy emerged.

However, this organisational approach does not necessarily fit creative work. In Chapter 2, we discussed the key factors in creativity as identified by Teresa Amabile and Tina Seelig—expertise and knowledge, creative thinking skills and imagination, intrinsic motivation and an appropriate attitude.

We also discussed how the process of creativity and innovation to solve problems and develop new products and services entails experimentation and failure, incubation and insight, challenging assumptions and the status quo, and exploring cross-domain combinations.

Most organisations, unfortunately, are not designed to facilitate and support the majority of these behaviours. By focussing on efficiency, reproducible processes, risk-management, and predictable and forecastable productivity and profit—all key considerations in running an efficient and profitable business—most organisations are actually set up to hinder creativity.

Those in charge cannot necessarily be expected to ignore the business requirements of their organisation and the need to bring in revenue. But by working towards the business goals of today, they may also be inadvertently designing and supporting organisations that systemically crush the creativity and innovation that will achieve the goals of tomorrow.

This chapter examines the conditions, behaviours and practices that are drivers of creative cultures within organisations. It provides a range of tools for stimulating innovation through careful leadership and management, improved structures and processes, effective team design and management, and better organisation design. Several checklists for self-reflection and/or use with groups and teams are included.

"Any company not willing to reinvent itself is doomed," said Fujio Cho, honorary chairman of Toyota, and an employee of the company since 1960 (3). Organisations must foster cultures that are open to new ideas, are agile and responsive to change, and are supportive of creativity and innovation if they are to survive for the long haul.

Creative Cultures

Consider how you greet someone you know when you meet them in the street. Do you shake hands with them, kiss them on the cheek (on one cheek or two…or even three or four times?), bow to them, hug them? What combination of those? Our culture will define what approach to greeting is appropriate, depending on who they are (business colleague or close friend?).

Getting it wrong can be awkward at best (shaking hands with a friend when a hug or kiss on the cheek might be more appropriate), but can also be highly insulting (a man hugging a woman who is not their wife or family member can be highly problematic in some cultures).

Culture is sometimes defined as 'the way we do things around here'. Or, as the dictionary puts it, "the customary beliefs, social forms, and material traits of a racial, religious, or social group" (4). Culture embodies and prescribes the way that people behave, what they believe to be important, how they see things as right or wrong, what they see as 'normal' and 'appropriate'. Contravening these cultural dictates can lead to sanctions of varying severity, from ridicule to incarceration.

Cultures within organisations are equally powerful. Interrupting or disagreeing with a senior staff member who is speaking in a meeting might be acceptable (and encouraged for the sake of dialogue) in some companies, and utterly taboo in others. Organisations, and their leaders, create cultures by virtue of the behaviours they demand, support, role model, and reward. And this culture influences all approaches an organisation takes to conducting its business.

There can be a considerable diversity in the culture of businesses. In the days when Ross Perot ran IBM, he insisted that everyone had to wear a white shirt and he would not tolerate beards. At their competitor Hewlett Packard, there was a very different set of cultural norms. In their early days, they were totally relaxed about what people wore to work, as they wanted to encourage people to express their individuality and stand up for their own ideas.

To foster and support creativity and innovation, therefore, organisations need to cultivate an organisational culture that allows, proscribes, and values the behaviours and norms that lead to creative outcomes and innovation. But the first step is to understand what those behaviours and norms actually are.

A number of researchers and academics have conducted significant research within organisations to ascertain the factors that have an impact on the creative outcomes of those organisations. In particular, they have sought to identify the dynamics and behaviours—particularly of leaders and managers—that establish an organisational culture that is conducive to creativity and innovation.

Goran Ekvall, professor emeritus of organisational psychology at the University of Lund in Sweden, spent a number of years looking at the organisational climatic factors that affected organisational creativity, and developed the *Innovation Climate Questionnaire* as a diagnostic tool for creativity within organisations. Its 10 dimensions are (5):

1. Challenge
    The extent to which employees are challenged, emotionally involved in, and committed to their work.
2. Freedom
    The extent to which employees are free to decide how to do their jobs.
3. Idea Time
    The extent to which employees have time to think things through before having to act.
4. Dynamism
    The eventfulness of life in the organisation.
5. Idea Support
    The availability of resources to give new ideas a try.
6. Trust and Openness
    The extent to which employees feel safe speaking their minds and offering different points of view.
7. Playfulness and Humour
    How relaxed the workplace is, and the extent to which it is ok to have fun.

8. Conflicts

    The extent to which people engage in interpersonal conflict or 'warfare' within the organisation.

9. Debates

    The extent to which people engage in lively debates about the issues within the organisation.

10. Risk-taking

    The extent to which it is okay to fail.

According to Ekvall's research, organisations that have these dynamics established as part of their culture will be more conducive to creative efforts and successful innovation. However, establishing these climatic conditions requires considerable work from all areas of the business.

Another team of researchers (6) identified a series of factors that directly correlate with the generation and implementation of new ideas. Their model is shown below. The key factors that have an impact are:

1. A systematic thinking style

    This is indicative of a disciplined approach to problem-solving (see the tools and approaches discussed in Chapters 5 and 7).

2. Role breadth self-efficacy

    This assesses the extent to which people have confidence in performing activities that are beyond the prescribed scope of their job.

3. Trust that benefit will accrue from the creative work

    There is an expectation that those managing the organisation have one's best interests at heart, and that creators will benefit in some way if an idea is implemented.

4. Trust in being heard

    There is an expectation that one's ideas will be listened to seriously.

5. Support for innovation

    This denotes the extent to which the organisation encourages the search for new ideas.

6. Leader member exchange

    This denotes the extent to which an individual's manager encourages involvement in the search for better ways of doing things.

These factors then influence two processes:

- The suggestion of ideas—the pursuit of new lines of thinking and the generation of new ideas.
- The implementation of ideas—the extent to which these ideas are moved forward and implemented.

The research concludes that, "the more an individual feels they are listened to and taken seriously, the more effort they put into having their suggestions implemented."

Factors affecting the innovative environment

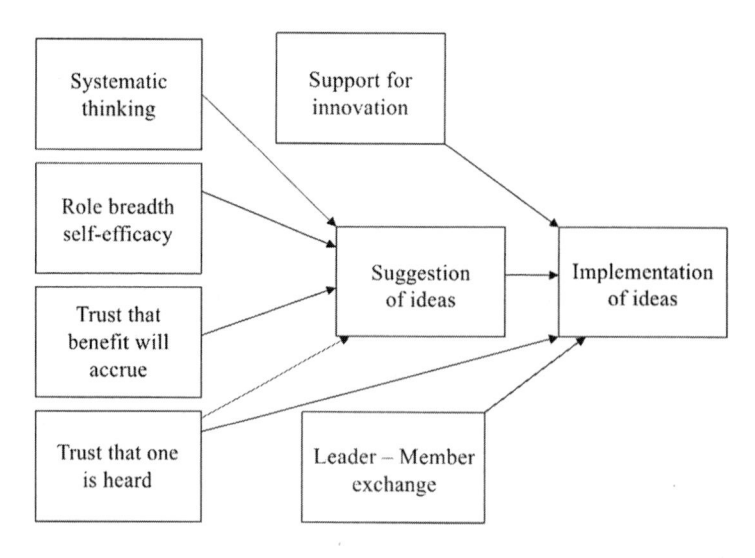

[Diagram: Systematic thinking, Support for innovation, Role breadth self-efficacy, Trust that benefit will accrue, Trust that one is heard, and Leader – Member exchange all feed into Suggestion of ideas, which leads to Implementation of ideas.]

Creating the right culture has many levers to bring it about. The organisation design is an initial building block, but of itself will not magically produce new ideas. Only when it is combined with shifts in management style, teams working with diverse talents, reward systems and resources of time and money that are appropriately deployed to explore and experiment, will the probability of innovation increase.

Apple is often cited as an example of an innovative organisation, and the company has embraced a number of principles that have driven their innovative culture.

Seek ideas from everywhere. Ideas can come from anywhere—not just within the organisation and its research and development departments—and the skill is in spotting them and mixing them together to create new products. It is easy to mistakenly assume that customers cannot envision new products.

However, it is common that innovation originates with the users of products realising these products don't quite work effectively in the situation they are in and dreaming of helpful changes, or when users experience situations that suggest a use for a product with a novel mix of capabilities (a gadget that mixes a phone, diary, email and music player, for example).

Be user- and consumer-driven and, with regard to technology, never underestimate the technical abilities of many consumers. Innovations that are successful—rather than merely interesting inventions—are about products that people actually want, and so understanding customers and their needs is important to implementation of new products.

Be aware, however, that there are some successful innovations that consumers never knew they wanted until they were made aware of the benefits—from spreadsheets (What is wrong with lined paper and a calculator?), mobile phones (Surely there are enough public phone boxes?) to VCR video players (But there are no videos to watch).

Design products for tomorrow and be resilient when today's users reject them. They may have too much emotional investment in the current state to truly appreciate and be open to new ideas. As a result, they must be gradually persuaded of the new products' benefits.

Learn from mistakes. Apple has had its fair share. (Remember the Lisa[15]?) But Apple has also been very successful at using their mistakes as learning opportunities and motivation to try again.

Pixar, the animation movie studio responsible for hit movies such as *Toy Story*, *Finding Nemo* and *Cars*, similarly has a series of directives that are designed to foster a creative culture, revolving around the following themes (7):

---

[15] Lisa was one of Apple's early personal computers, and a commercial failure.

- Hire people smarter than you.
- Fail early, fail often—failure "is a necessary consequence of doing something new".
- Listen to everyone's ideas—"inspiration can, and does, come from anywhere".
- Face towards the problems—"engaging with exceptionally hard problems forces us to think differently".
- B-level work is bad for your soul.
- It's more important to invest in good people than in good ideas—"if you get the team right, chances are that they'll get the ideas right".

Accordingly, we will discuss a number of important considerations in the following areas:

- Leadership for creative cultures
- Job and organisation design
- Innovative teams
- Resources and rewards
- Architecture and creativity

Leadership

Psychologists Zhou and Shalley (8) conducted a comprehensive examination of organisational creativity, and noted—perhaps not surprisingly—that the way individuals and groups are managed had a considerable impact on the effectiveness of creative organisations.

Leadership of an organisation, and the management of teams and individuals within it, set the tone for creative activity and the resulting innovation. Not only must leaders and managers carry out their roles in ways that establish, support, and champion a creative culture, but they must also grant permission—explicitly or implicitly—to staff in all corners of the organisation to act creatively in the first place. And this is easier said than done.

Assessing effective leadership behaviour can be challenging. Like a black hole that is only discovered by the impact on its surrounding environment, so the result of leadership behaviour can only be seen by the impact it has on individuals and teams within the organisation.

A crude test of this impact is to assess how people feel at the end of the working day. Are they exhausted, frustrated, resentful? Or tired but energised, challenged but motivated to keep working at a problem, supportive of the vision and mission?

Much has been written about the theories and practices of effective leadership, so we will restrict our discussion here to the elements of leadership and management that can have an impact on creativity within organisations. Here are some guidelines for leaders to consider.

1. Act as a role model for creativity.

Champion creativity from the top down. How leaders behave tends to set the tone for how others behave, so leaders should aim to act in authentic ways that promote and pursue creative outcomes, and support and promote the process of innovation. As we've discussed throughout this book, creative ideas stem from experimentation, a willingness to fail and to learn from that failure, and perseverance.

Leaders must participate in the process and role model this behaviour to the rest of the organisation. In doing so, they must be prepared to accept the failure and fallibility that come with these creative endeavours.

Research by renowned Harvard Business School professor Clayton Christensen (9) found that organisations with higher rates of innovation had leaders with a track record of innovation themselves, and that those leaders saw it as their responsibility to lead the process of innovation from the front.

These leaders also displayed 'discovery skills' that included the capacity to connect seemingly unrelated concepts, and to question and challenge the status quo rather than falling into the trap of assuming the current way of doing things to be the only way. They also demonstrated a willingness to experiment and seek ideas from a diverse range of sources.

Some further suggestions:

- Adopt a style of leadership that is prepared to experiment with new ideas, and is not afraid to admit when it is wrong.
- Make creative suggestions and introduce new ideas without being overly dominant (avoid an attitude of 'my way or the highway').
- Role model a degree of risk-taking that includes pursuing unconventional approaches and challenging existing rules and policies.
- Encourage creative thinking and the use of idea generation techniques.
- Demonstrate a commitment to persevere when setbacks arise rather than giving up easily.
- Keep curious and seek opportunities to continue learning.

2. *Share ideas and encourage information flow; recognise that good ideas can come from anywhere.*

Leaders must positively encourage the organisation to share knowledge and collaborate across organisational boundaries. Information flow is key to creative success, ensuring that knowledge and expertise—from market intelligence to research and design strategy—flow to and from all corners of the organisation, where everyone can make use of them as the raw materials of creative endeavour.

In particular, communication between people in different areas of the business, or those that have different perspectives on the business, can lead to the unexpected combinations or the reapplication of ideas and information that lead to creativity and innovation.

Unfortunately, information gets stuck within teams and organisational silos. Teams resist collaboration and people hold on to information as a means to broker power within an organisation. This is ultimately destructive and undermines the organisation's capacity for creativity and innovation, given the importance of both knowledge/expertise, and also diversity of perspective in the creative process.

Since no organisation design is ever perfect, making the structures and processes work as efficiently as possible is vital. This also means reducing the politics and turf wars that can arise. The task of leadership (10) is to reduce conflict and disequilibrium—not in order to suppress legitimate debate, which is always important in developing new ideas, but in order to reduce the corrosive and destructive nature of power politics.

The Post-it Note has become a common stationary item in offices around the world—a small piece of paper with a re-stickable strip of glue on its back, made for temporarily attaching notes to documents and other surfaces. In the early 1970s, Art Fry, who worked in product development for the 3M manufacturing company, was in search of a bookmark for his church hymnal that would neither fall out nor damage the hymnal.

Fry noticed that a colleague at 3M, Dr Spencer Silver, had developed an adhesive in 1968 that was strong enough to stick to surfaces, but left no residue after removal and could be repositioned. Fry took some of Silver's adhesive and applied it along the edge of a piece of paper. His church hymnal problem was solved.

Fry soon realised that his 'bookmark' had other potential functions when he used it to leave a note on a work file, and co-workers kept dropping by, seeking 'bookmarks' for their offices. This 'bookmark' was a new way to communicate and to organise.

3M crafted the name *Post-it Note* for Arthur Fry's new bookmark and began production in the late 70s for commercial use. In 1977, test markets failed to show consumer interest. However, in 1979, 3M implemented a massive consumer sampling strategy, and the Post-it Note took off. The rest, as they say, is history.

Some further suggestions:

- Actively encourage and solicit ideas from all over the company.
- Recognise that seniority confers no special powers in developing new ideas and that the only way to innovate is to listen and learn from others, and to experiment.
- Encourage trust and openness with information and ideas.
- Employ an information management system to collect and manage ideas, suggestions and feedback.

3. *Stimulate debate and encourage questioning and curiosity; tolerate ambiguity, disagreement and even dissent.*

Information flow isn't just about facts and figures. The ideas, perspectives, and viewpoints held by every staff member also hold great value. Furthermore, when these various ideas and perspectives can meet in open debate, creativity and innovation can ensue. Assumptions are challenged and ideas are born through dialogue.

Leaders should therefore seek to stimulate dialogue and debate amongst staff, and encourage them to question the status quo, to question assumptions, and to ask the key questions, 'why?' and 'what if?'

In the mediaeval court, the jester was given permission to say the things that no one else dared say. In modern organisations, we rarely have such a role, so the permission to speak up must be embedded in the culture.

When organisations are stuck rigidly in their patterns of doing and being, without ever questioning them, the capacity for creativity is greatly diminished. Furthermore, organisations run the risk of 'group think', where everyone subscribes to a prevailing perspective (usually emanating from the top of the organisation) and no one dares question it.

At its worst, group think can be fatal—as it was in the case of the Challenger shuttle disaster, when the warnings of engineers, who were aware of issues with the O-ring seal that likely caused the disaster, were unheeded as NASA was committed to pressing ahead with the launch.

At a lesser extreme, group think undermines an organisation's capacity to entertain diverse perspectives, to question assumptions, to tease out fatal flaws, and to make unexpected combinations leading to innovation.

The answers to the question, 'what can we not talk about?', are indicators of how much freedom individuals have to challenge the status quo and the organisation's sacred cows.

Such debate and questioning will, of course, lead to disagreement and the challenging of beliefs, norms, behaviours that might have been instituted at the very top of the organisation. But this must be given permission to happen, and leaders must be prepared to tolerate dissent if it is in the service of achieving the company's goals or innovative breakthroughs.

This is not to say that the ends justify the means, and that any form of behaviour that undermines values and breaks rules or even laws can be justified if it leads to successful outcomes. However, to provide and establish a safe space for people to share these diverse perspectives in dialogue, it must also be safe to disagree, to question one another's ideas, and to question any assumptions and cultural norms that might be actively hindering creative endeavour.

A moment of truth for encouraging innovation is when rules are broken. Will people be forgiven in the interests of experimentation and learning? Or will there be punishment?

A British airline wanted to encourage greater flexibility amongst staff who were involved in meeting the arriving and departing aircraft, in particular the ramp handling staff who looked after the maintenance and refuelling at the stand. Ramp handling staff had a tough manual job with plenty of rules to be followed, and they received a training programme designed to empower them to make more local decisions.

A little later, a ramp supervisor heard that a busload of students was stuck in traffic and was in danger of missing a domestic flight from London to Newcastle. Recognising that take-off slots were expensive if missed, motivated by a desire to help the customers and feeling that he was empowered to do so, the supervisor ordered the bus to drive straight up to the steps of the aircraft, much to the delight of the students.

The plane was able to take off on time. However, the supervisor had broken numerous security rules and was disciplined as a result. All the other supervisors saw this and realised that nothing had really changed, despite the training programme, and that the rules still had to be obeyed no matter what.

When the airline gave the situation further consideration, they realised that they should have gathered everyone together and praised the supervisor for having a go and thinking creatively to solve the problem. Then, with everyone there, they should have had a discussion about how such scenarios could be handled in other creative ways without breaking the rules.

That way, the supervisors would have felt positive about the greater trust being placed in them, and also have learned some tactics to better handle such difficult situations in the future.

Some further suggestions:

- Make it safe for people to speak their minds and contribute differing perspectives and points of view.
- Hold a space for conflict, as long as it is constructive rather than destructive.
- Encourage lively debate and intellectual stimulation.
- Encourage the use of devil's advocate and 'red teaming' (a technique used extensively in the military to assess everything that could go wrong and ensure a counter to an overly positive view) in debate.
- Avoid rewarding employees who are over-critical (as if this signals competence and intelligence).

4. *Mix freedom and autonomy with clear direction.*

Despite the modern architecture in which they operate, many workplace cultures are still predominantly based on an outdated assumption about work, namely that it is conducted at a desk or workbench between specific times of the day. Ideas, though, are not so constrained, and organisations must experiment with giving greater freedom to staff to undertake their work and pursue their ideas in the ways that suit them best.

The key is to clearly define strategic and organisational goals, and then give employees the freedom and autonomy to determine how to achieve them. Leaders must make sure the mission and vision of the organisation are clear, and should work hard to ensure that employees share in the same enthusiasm for them.

Tell people *what* you want them to achieve, and let them figure out *how*. Doing this gives people a sense of autonomy—one of the key ingredients of intrinsic motivation—as well as a sense of ownership and control over the process, and this sense of freedom and autonomy as people go about their jobs, is very powerful in maximising intrinsic motivation and creative outcomes.

Granting autonomy to employees can often be challenging, particularly in global organisations. If a digital strategy for a simultaneous global release needs to be coordinated across territories (for the new Beyoncé album for example), granting certain freedoms just might not be possible.

But such freedom allows people to approach challenges in ways that make the most of their expertise and their creative thinking skills—the strengths with which they can meet the challenges of the roles, tasks and responsibilities they've been given.

Furthermore, as we discussed in Chapter 3, everyone has individual differences in how they perform at their best, and the more freedom people are given to set up working conditions that are in line with the conditions in which they are at their most creative, the more the organisation will benefit.

These freedoms do, of course, actually need to be supported throughout the organisation. It's no good telling people that they have some freedom in the working hours that they choose, and then having others (particularly senior staff) chastise or make fun of them for leaving the office at 4 pm ('leaving early?' 'half day?'), particularly if they've been working since 6 am.

If employees do not feel safe to exercise their freedom, and run the risk of criticism or punishment for diverting from certain norms, then this quest for freedom and autonomy becomes a pointless exercise.

Such freedom is a cornerstone of animation movie studio Pixar's approach to fostering creativity. As Ed Catmull, Pixar's founder and president, reflected, "You get great creative people, you bet big on them, you give them enormous leeway and support, and you provide them with an environment in which they can get honest feedback from everyone" (11).

However, it is a difficult task to communicate your vision down the organisation in a meaningful way. A journalist listened one day to a presentation by the chief executive of a hotel chain, which described the chain's business strategy in complex and detailed fashion.

The journalist wondered how that strategy, articulated in such sophisticated language, became real for the people at a junior level in the organisation. He went into the bar of the nearest hotel in this chain and asked the barman, "What's the strategy of your company and what is your part in it?"

He rather expected a blank look, but to his surprise, the barman said, "Well sir, we are trying to attract a more sophisticated clientele and my job is to make a range of interesting cocktails."

This demonstrated that despite the sophisticated language used in the presentation, the hotel chain's senior management had done an effective job in articulating the chain's vision and strategies throughout the organisation. Asking such questions of staff is a good test for any manager.

Some further suggestions:

- Set a clear and challenging vision and strategy, and give employees freedom and autonomy to respond in a way that works best for them.
- Ensure goals and targets are clear, consistent and challenging but reasonable—vague, ever-changing, or impossible goals will undermine the process.
- Avoid constantly changing direction with little clarity about what is expected, or focussing on routine and past practices.
- Trust staff to do good work, with minimal rules in place.
- Avoid focussing on controlling behaviour according to the rules.
- Encourage people to care for the organisation's vision and mission.

5. *Grant freedom to try new things, including the development of competing options.*

Successful leaders in creative organisations recognise that a powerful approach to harnessing an employee's intrinsic motivation in the quest for creative ideas is to let them pursue their own ideas and hunches, even if they fall outside the scope of their jobs.

In a TED talk (12), author Steven Johnson discussed the genesis of GPS. Two scientists at a lab were discussing the Sputnik satellite that had been recently launched and was orbiting the earth.

Out of curiosity, they wondered if they could listen for its signal and track it. They quickly found its transmission frequency—apparently, the satellite broadcast on a very common frequency so that everyone would be aware of its presence and know it wasn't a hoax—and, using a knowledge of the Doppler effect, were able to calculate its whereabouts.

Through a series of discussions, they then reversed the maths and realised that if they knew where the satellite was in space, they could then figure out where something was on earth. And this was how GPS was formed. This only happened because they had:

- the curiosity to ask the questions
- the knowledge and expertise to know how to explore the answers
- the resources to track the frequency, and the time to experiment

- the organisational support by way of their boss who shared their enthusiasm and curiosity

This implies that there is some latitude in budget and time to do more than is prescribed by the needs of the business today. Accordingly, organisations must strive to provide resources—time, budgets, etc.—for staff to try new things and pursue new ideas. Leaders must be prepared to take the risk of allocating resources away from business as usual (see Chapter 8).

In an approach to embedding innovation at the 3M manufacturing company, scientists were expected to devote 15% of their time to working on new ideas that were not part of their work schedule. This approach to time for employees to dedicate to individual pet projects has been implemented at other organisations, sometimes known as '20% time', whereby employees can dedicate 20% of their time to working on these pet projects.

It can be very challenging to protect this '20% time' from other organisational demands, however. Google has talked about a balance between the main job (about 70% of the time), core pursuits and pet projects determined by the individual (20% of the time), and far-out ideas (10% of the time).

Software company Atlassian formalised this freedom to experiment in their infamous concept of the 'ShipIt Day'—a day once a quarter when you can work on whatever you want (whatever idea inspires you), you can assemble your crew to work on it with you, and then you have 24 hours to 'deliver' it (hence the ShipIt moniker, also referred to in other contexts as FedEx days). An explanation on the company's website reads, "24 hours to innovate. It's like 20%. On steroids."

A similar approach, called 'skunk works', entails the formation of a group within an organisation that is given a high degree of autonomy, is unhampered by the demands of bureaucracy, and is given the task of working on advanced or secret projects.

This approach is widely credited to Lockheed Martin aerospace—where the approach was christened—with the development of several new aircraft, including the U2 and Blackbird spy planes. The 'new ventures' group at consumer goods manufacturer Proctor and Gamble has also enjoyed great success in accelerating the development of new products within the organisation.

The danger of these separate groups, however, is that they can create the belief throughout the organisation that those who are not part of the group are absolved of pursuing and developing new ideas.

Leaders should aim to inspire and excite staff with possibility rather than crush them with constraint. "I never tell people what to do but rather, help them see the possibilities, let them really get excited about one of them, and let them work on their own ideas," said Professor Gordana Vunjak-Novakovic, who leads a team of 40 in a medical research lab at Columbia University (13).

Some further suggestions:

- Use specific dialogue to encourage the development of ideas (see Chapter 3).
- Empower staff to explore their own interests and pet projects.
- Avoid insistence on heavily prescribed roles and sanctions for stepping outside those roles.
- Avoid meeting ideas with long delays and critical evaluations, or looking for reasons why an idea will not work.

6. *Recognise that rules and procedures were the answer to yesterday's problems but the creators of today's.*

A mindset that follows the rules or is a slave to the logic of any solution will inhibit creative thinking. Organisational structure assumes a way of organising that can inhibit new ways of working, encouraging a 'silo mentality'. Fluidity in ways of organising and physically locating people can reduce these limitations.

Assume that 'the cement is always wet' and that all structures can be altered. Of course, the rules might be eminently sensible and necessary; it's when blind acceptance of these rules becomes the norm that problems arise.

Some further suggestions:

- Be flexible and prepared to adapt as new ideas emerge; adopt the mindset that you are always learning.
- Don't rely on only using tried and tested approaches. Avoid resisting change until there is no choice, then changing priorities frantically.

- Place less focus on existing ways of doing things, and avoid engaging in turf wars and politics to protect existing ways.
- Avoid discouraging anything outside of the current rules and practices.

7. Encourage and support diversity.

Diversity is a pertinent issue for many organisations to consider, driven largely by political considerations around equality of opportunity for different groups (e.g., gender, ethnicity, sexual orientation, faith, etc.). However, as we will discuss in greater detail later in the chapter, various studies have also pointed to the economic benefits of diversity within teams and organisations, as it is a key driver of creative thinking and innovation within those teams and organisations (14).

For example, consumer goods company Proctor and Gamble identified real improved performance in situations where a diversity of gender, race, culture, age, mother tongue, sexual orientation, faith, education or personality enabled a greater variety of thinking styles and experiences. This diversity sparked challenging debates between differing view-points, and fostered learning through seeing differences in values and priorities.

This diversity, however, can be challenging to manage successfully, and can come at a cost. Differences trigger wariness and so it can take longer for teams to build the trust that is a prime requisite to work well together, as well as give discretionary effort and share ideas, especially if their ideas are off the wall.

Diverse teams can therefore have a great variance in their performance—for better and for worse—and must be carefully assembled and managed to be productive. We discuss this in greater detail shortly.

Some further suggestions:

- Balance diversity in a team with the need to ensure effective interaction.
- Champion an organisational commitment to diversity.
- Firmly reject any processes, practices or behaviours that lead to discrimination against diversity, or undermine access to diversity.
- Proactively seek out access to diverse perspectives and individuals.
- Consider and challenge any biases that narrow diversity (e.g., where recruitment notices are placed, etc.) (15).

8. *Consider the necessity and value of failure.*

Leaders must carefully consider how they respond to failure—on an individual level and on an organisational level. Whilst success is the lifeblood of an organisation's existence, it's today's failures along the way that provide the learning and the springboard for tomorrow's success. As Yoda tells any aspiring Jedi in the *Star Wars* movies, "The greatest teacher, failure is."

Accordingly, leaders should be sparing with criticism when an experiment goes wrong, treating it as a potential learning experience that allows people to try again more intelligently. Rather than sweeping failure under the carpet and swiftly moving on, a post-mortem of a failed effort, project, or experiment can be insightful in producing the learnings that lead to future success. Such an effective post-mortem is only possible if the shame of failure is reduced, and people are prepared to sit with, and appraise the details of, their failure.

When Thomas Edison was asked why he wasn't getting results with his countless attempts to successfully develop the light bulb, he replied, "Results? Why, I've got a lot of results. I know several things that won't work" (16). Failure is a great teacher and is often instrumental in the breakthrough to new products and services.

Some further suggestions:

- Use positive encouragement to keep pushing on when things go wrong.
- Celebrate creative efforts.
- Avoid criticising failure and apportioning blame.
- Be curious to understand how to do better.

9. *Recruit for creativity and know your staff.*

How can you recruit future generations of staff who will contribute in powerful ways to the organisation's quest for innovation?

Taylor and LaBarre (17) pose the following pertinent questions: Why should great people join your organisation? Do you know a great person when you see one? Can you find great people who aren't looking for you? Are you adept at teaching great people how your organisation works and wins? Does your organisation work as distinctively as it competes?

Since creativity is enhanced by effective teamwork, a critical management task is to get the right mix of innovative thinkers and adaptors within the organisation, and to build effective relationships to turn ideas into practical products and services.

Recruitment for creative people is based not just on what they have done but how they think and the likelihood that there will be a synergistic relationship with others. Another key factor is how they respond to failure and criticism. Do they display those qualities of grit and resilience we have discussed in earlier chapters?

Furthermore, leaders must recognise that leadership operates in different ways at different levels. At the group level there is a greater focus on the use of facilitation skills to encourage synergy; however, at the individual level, the degree of personal relationship is important because it impacts trust.

For leaders, these personal relationships are important not only with one's direct reports, but also with employees at every level of the organisation.

People want to feel a sense of relatedness—to be part of a team or community—and a personal relationship with their leaders and managers can be extremely powerful. When you consider your staff, do you know their birthday? Their favourite sport or sporting team? Or musician/artist? Whether they have kids? Leaders who have demonstrated this level of knowledge about their staff have fostered powerful cultures of trust and relatedness, that can drive improved performance and innovation.

Finally, an important aspect of knowing your staff is to match them in teams and with roles accordingly. Not only should jobs and roles be tailored to the individual employee by giving them freedom to choose and define their activities, the job should be crafted to challenge the employee in ways that play to their strengths, interests and expertise.

Creative cultures flourish when employees are matched with roles that challenge them enough to keep them motivated and engaged. This match of person to job in a way that maximises their strengths is an important element of creativity.

A good match stretches the employee's ability enough to present them with a challenge, a task to achieve, a process to master. It does not challenge them so little as to leave them bored, uninspired, and disengaged (and probably wanting to leave the organisation), nor too much to overwhelm them and make them feel stressed out. Organisational support may be key here to ensure that burn out does not take place.

Some further suggestions:

- Be emotionally connected to your people. Display positive personal interaction, empathy, curiosity, and humility.
- Avoid being impersonal and transactional, dispassionate and disconnected.
- Be prepared to hire outside the conventional staff profile.
- Match employees with assignments appropriate to their skills and interests.
- Avoid placing people in jobs regardless of their skills, interests, and wishes.

10. *Give effective feedback.*

One of the hardest skills for any manager to learn is how to give feedback, but feedback is one of the most powerful ways that organisational performance can be guided.

It is through both formal appraisal type schemes and informal mechanisms that individuals learn what is acceptable. Such feedback tends towards normative behaviour and hence is inhibiting of truly creative work. It is often confused with criticism.

Criticism is often taken personally and leads to wariness and defensiveness, whereas feedback deliberately tries to encourage the person to keep trying to improve but with more insight into how they are doing. In developing new products, feedback is essential to learning, but, done badly, can kill the will to keep working at the project.

Consider also how the prevailing culture can colour the interpretation of feedback, which in turn can amplify the cultural attitude towards feedback. In the table below, the different interpretations of familiar phrases depend on whether prior experience has led to a belief that there is a 'blame' culture or a 'forgiveness' culture. Clearly, a blame culture means that feedback can be interpreted in destructive ways.

The influence of culture

	Blame Culture	Forgiveness Culture
"You didn't do that very well."	Accusation	Opinion
"I hope you manage better next time."	Threat	Encouragement

Used with permission from Rita McGee.

As a foundation for effective feedback, remember that:

- Feedback is given with positive intent, to help the recipient improve. Clarify the benefits of receiving this feedback as you are giving it.
- Remember to focus on the facts of what happened, using specific (and preferably first-hand) examples.
- Explain the impact that what happened (and the behaviour that led to it) has had—on the organisation, on you or others.
- Consider how to discuss actions moving forward that can make change for the best based on this feedback.

Curiosity can be a key part of the process as the person giving feedback employs curiosity to truly understand the perspective of the recipient, rather than just making assumptions about what was going on for them, and wading into the feedback conversation with judgement.

The question 'why did you do that' can be a powerful tool to dig into something, uncovering hidden assumptions and pushing towards new understandings. However, it can come across as doubting the person's ability if the tone of voice is perceived as critical.

We examine feedback in further detail in Chapter 7.

The Leadership Questionnaire

This checklist can be used diagnostically with individuals or groups of managers to encourage reflection about their roles and whether the balance is correct for what the organisation needs to achieve. Feedback from subordinates might be helpful to check the accuracy of the manager's self-perception.

You can rate yourself on each variable using the 1-5 scale, whereby 1 is a strongly helpful style and 5 is a strongly hindering style. Then assess what possible action you might be able to take in response to your self-assessment.

Helpful style	Rating 1 2 3 4 5	Hindering style	Action
Making creative suggestions and introducing new ideas without being overly dominant.		'My way or the highway.'	
Setting a clear and challenging future vision and strategy, and giving employees freedom and autonomy to respond in a way that works best for them.		Constantly changing direction with little clarity about what is expected. Focussing on routine and past practices.	
Emotionally connected to their people. Displaying positive personal interaction.		Impersonal and transactional, dispassionate and disconnected.	
Encouraging people to care for the organisation's vision and mission. Trusting staff with minimal rules in place.		Focussed on controlling behaviour according to the rules. Overly financially focussed at the outset of projects.	
Not overly protective of their vision if circumstances change.		Resistant to any challenge to their vision and way of doing things.	
Flexible and prepared to adapt as new ideas emerge. Always learning.		Only using tried and tested approaches. Resisting change until there is no choice, then changing priorities frantically.	
Placing less focus on existing ways of doing things.		Engaging in turf wars and politics to protect existing ways.	

Looking for new ideas from every source, particularly customers.		New ideas are the sole responsibility of one department (e.g., marketing).
Encouraging development of ideas, using encouraging dialogue (Chapter 3).		Being overly critical of new ideas and shutting them down.
Self-starting—not necessarily waiting to follow others' lead.		Hierarchy is paramount and decision-making is slow.
Persevering through setbacks.		Giving up easily.
Being prepared to take risks by allocating resources away from business as usual (see Chapter 8).		Over- or under-resourced, including inappropriate space and equipment.
Capable of envisioning the future with a particular innovation in place, and the challenges along the way. Possessing the courage to make a 'go' decision.		Having short term view blocked by perceived barriers that they believe are insurmountable.
Role modelling of risk taking, pursuing unconventional approaches, and challenging existing rules and policies. Encouraging lively debate and intellectual stimulation.		Discouraging anything outside of the current rules and practices. Rewarding employees who are over-critical (as if this signals competence and intelligence), and punishing failure.
Being prepared to hire outside the conventional staff profile.		Unwilling to hire people with different experience and character that complement existing teams.

Balancing diversity in a team with the need to ensure effective interaction.	Creating teams whose diversity becomes a source of division.
Being prepared to create unconventional working environments.	Insisting on staid, fixed and traditional office environments.
Matching employees with assignments appropriate to their skills and interests.	Placing people in jobs regardless of their skills, interests, and wishes.
Empowering staff to explore their own interests and pet projects.	Insisting on heavily prescribed roles and sanctions for stepping outside those roles.
Setting clear, reasonable, but challenging goals.	Failing to clarify goals, setting impossible goals, setting fake deadlines.
Sharing power.	Hoarding power.
Using positive encouragement to keep pushing on when things go wrong. Avoiding blame, and being curious to understand how to do better.	Criticising failure and apportioning blame. Meeting ideas with long delays and critical evaluations. Looking for reasons why an idea will not work.

Organisation Design – Job Design, Structure and Process

Organisations want to ensure that an innovation—a product or service that they have successfully introduced to the market—continues to thrive beyond its launch. As a result, once the initial development and launch stage has passed, organisations tend towards creating stable and reproducible processes, certainly as far as that product or service is concerned.

The gadget, the car, the hotel room, has to be reproduced many times if the business is to prosper, and if it can be reproduced with increasing efficiency, then business will prosper even further.

As a result, the traditional structures of organisation—division of labour and hierarchy—are designed for the reproducible and consistent world. Innovation, which will drive the success of tomorrow, requires a breaking of these patterns.

Ignoring innovation and doubling down on the reproducible product puts a business at great risk, and history is littered with organisations who focussed too heavily on their current success (and rested on their laurels), at the expense of creative approaches and the quest for new innovation. The consequence can be that another business introduces an innovation to the market instead, immediately killing demand for the first organisation's product.

Kodak is a classic example of a company that focussed too heavily on its current area of success (in camera film), and was completely unprepared for the revolution of digital photography as a result.

Organisations must therefore encompass two different, and often opposing, strategies: the continuing success of existing products and services in the most efficient way, and the development and launch of new products and services. The former process entails stability, to ensure that best practice is retained and honed to ensure consistent delivery. In this situation, improvement comes from working with existing products, processes, and (invariably) people over a period of time.

By contrast, the process of developing and launching new products and services—the innovation process—entails a great deal less stability, as employees are given increasing freedom to experiment and do things in a way that suits them, but this then increases the variability of performance.

Whilst stability and consistency may be detrimental to the more adventurous process of finding something new, it is consistent delivery of existing products and services that provides the rationale and delivers the income for more radical innovation to occur. So the two strategies must co-exist side-by-side, no matter how much friction this causes. Managing these two critical forces poses a challenge for all organisations!

Stability, best practice and consistent delivery often lead to clearly defined and proscribed structures and processes, which, while maintaining efficiency, can also undermine a capacity to be agile and to react to change.

As an example, Gillian Tett (18) cites Sony's failure to capitalise on the ubiquity of the Walkman and transition into digital media and the growth of MP3s as a means to consume recorded music. The organisation was quick to respond to the release of the first MP3 player; however, their own attempts at cornering the MP3 player market were undone by the lack of a coherent strategy.

Rather than uniting under a shared vision, the different divisions within Sony Corporation were too interested in preserving their turf and therefore in producing rival products: as a result, Sony simultaneously unveiled three competing products. Furthermore, the organisation's recorded music arm, Sony Music, was more concerned with protecting revenue from CDs (as opposed to digital downloads) and fighting the piracy that MP3s permitted.

As a result, Sony completely missed the opportunity to dominate the MP3 market, and left the door wide open for Apple and its iPod.

Symbiotic with the silo effect is job design that often limits the freedom for employees to go outside of one particular specialism—a situation that may well be exacerbated by the increasing trends in specialisation within educational systems.

The following tactics can be useful to set up job design and organisational structures more effectively for creativity. These tactics also align with the leadership principles we have just discussed in the preceding section.

Team flex

Most jobs have core demands, the activities and outcomes that must be fulfilled if the job is to be satisfactorily performed to the minimum acceptable level. Beyond those core demands are other more optional activities that often depend on the experience, talents and inclinations of the job holder if they are to be done at all; the job holder can effectively choose which of these optional activities they do.

The more senior the job (Job B in the first diagram below), the more choice can be exercised in the make-up of the job. The shape of a job can differ therefore between two people who are supposedly doing the same job or role. The second diagram below, humorously described as the 'fried egg' model of job design, illustrates this flexibility.

Demands, constraints and choices

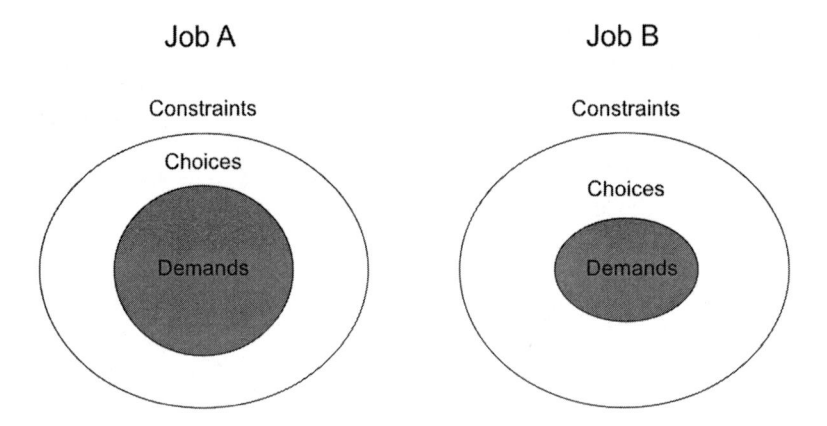

Job A

Constraints
Choices
Demands

Job B

Constraints
Choices
Demands

The 'fried egg' model of job design

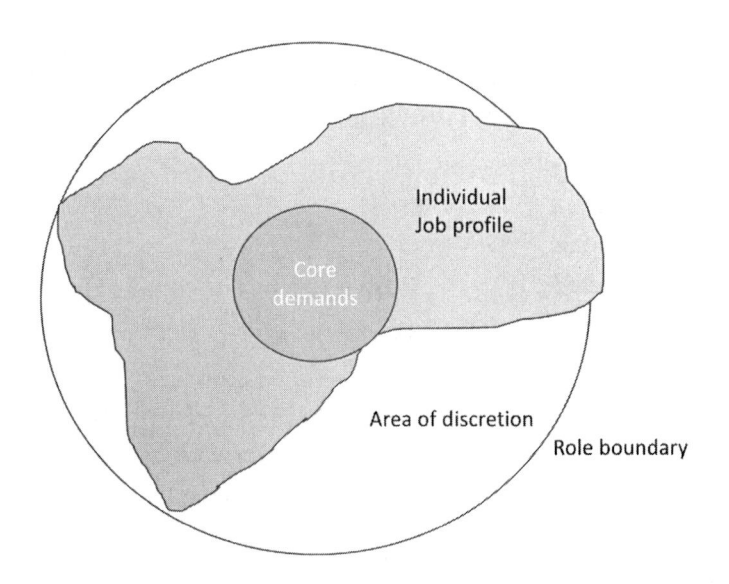

Individual Job profile
Core demands
Area of discretion
Role boundary

Team flex recognises that if we are flexible about the boundaries between jobs, there is a greater likelihood that people will be able to play to their strengths, interests, intrinsic motivations, and so be more motivated to contribute. People want to do good work—it's pretty rare that people go to work wanting to do a poor job!

So the more people can flex their jobs to play to their expertise, use their skills in creative thinking, and—arguably most importantly—ignite their intrinsic motivation, the more they are able to do a good job and exercise their creativity.

In the early days of Microsoft, a new recruit was hired to a team, but not to a specific role, and told to 'contribute'. At the end of the month, the rest of the team voted on whether the new recruit should stay.

Additionally, team flex ensures that there are less turf wars within the team itself.

Buddy system

A further development of the team flex process is the buddy system. At its simplest, this entails a colleague who can help out in dealing with surges of workload, or as a sounding board when problem-solving or generating ideas. This becomes more difficult as organisations grow in size.

There is a story about Bill Gore, co-founder of the W.L. Gore clothing technology company (and inventors of Gore-Tex waterproof material) walking on to the floor of his factory one day and realising he didn't recognise many people. Something told him this couldn't be good for him, his employees or the company, and he concluded that to maintain the sense of camaraderie, a factory should have only about 150 people (19).

Natural work teams

When problems arise between teams, there is a process to resolve them called 'natural work teams'. These are groups, often temporary, that are brought together to solve a problem, establish effective processes, or to pioneer some other innovation. These teams often consist of participants from a range of different departments, and can include anyone who has any impact on (or is impacted by) both the problem and the possible solutions.

Since team members are impacted by the issue, they are generally motivated to find solutions. By bringing together participants from a number of different departments, a variety of perspectives on the issue are brought to the table, and this diversity can fuel more effective and innovative solutions.

Natural work teams must be carefully facilitated, using a variety of tools, as we explore in Chapters 5 and 7.

Freedom and autonomy

As we have discussed, many workplaces are still predominantly based on the outdated assumption that work is solely conducted at a desk or workbench at specific times of the day. Since ideas are not so constrained, some organisations are starting to experiment with giving greater freedom to people to undertake their work and pursue their ideas in the ways that suit them best.

Freedoms to consider might include: location of work (work from home vs work from office vs work from café), hours of work (7 am-4 pm or 10 am-7 pm, for example), what tools to use, etc. Note that many of these freedoms that can be granted align with elements of increasing personal creativity that we discussed in Chapter 3.

Freedom about the process allows people to approach challenges in ways that make the most of their expertise and their creative thinking skills—the strengths with which they can meet the challenges of the roles, tasks and responsibilities they've been given.

Building on this freedom and the notion of team flex, some organisations recognise that a powerful approach to harnessing an employee's intrinsic motivation in the quest for creative ideas is to let them pursue their own ideas and hunches, even if they fall outside the scope of their jobs.

As we've discussed, not only should jobs and roles be tailored to the individual employee by giving them freedom to choose and define their activities, the job should be crafted to appropriately challenge the employee. Organisational support may be key here to ensure that being overwhelmed or burn out does not take place.

So, if you want people to be creative, place them in roles that challenge them in areas where they already have expertise, interest, and motivation.

Embed innovation in the DNA of the organisation's employees

The challenge many organisations face is how to embed innovation in the DNA of the organisation and its employees.

An organisation's culture must reflect the fact that good ideas can come from anywhere and anyone within the organisation. Employees must believe they have the capacity for creative ideas, and must be encouraged to share and discuss any ideas that they might have. The commitment to '20% time' for employees to explore their creative interests and ideas, as we discussed earlier, can be powerful here.

Further efforts to embed innovation in an organisation's employees can be employed as part of a leadership development strategy. One organisation had a leadership development programme with an organisation development agenda. The next generation of leaders were carefully chosen and trained together in cohorts.

At the end of two weeks of training, each cohort became a project team to work on a specific problem that a senior manager was currently wrestling with. This was an exercise in creating bonds amongst the next generation of leaders, so that they might be less concerned about their own departments and, with stronger relationships in place, more likely to collaborate across the organisation.

Furthermore, these project teams then contributed a number of original ideas that were adopted by the organisation. As one senior manager said, "I had become too close to the problem. The project team came at it with a clean sheet of paper."

Change the project manager's role

Organisations with project-based structures have experimented with variations in the project manager's role. Traditionally, project managers were responsible for managing both the project as a whole, and the staff within the project-based team. This broad responsibility could undermine the project manager's capacity to execute either task effectively.

Another approach gives the project manager responsibility for delivering solutions to time, cost, and quality expectations, whilst a separate pool manager is given responsibility for team staffing, with a KPI (key performance indicator) of supplying staff when needed by the project but also of maintaining maximum utilisation of staff members.

This frees up the project manager to concentrate on the project, whilst allowing the pool manager to focus on developing and retaining the best mix of skills to deal with different projects at different phases of the project.

Build flexibility into organisation design

Building flexibility into organisation design can be encouraged by several elements:

- Establish broad pay bands, where pay distinctions are reduced, thereby making it easier for people to move jobs or add or subtract duties without causing pay disputes.
- Remove layers of the organisational hierarchy to facilitate faster decision making and increased discretion among employees to pursue and experiment with ideas.
- Establish a physical layout of desks and offices that makes it easier to change teams and regroup functions, whether with an open plan office, or an internal office structure that can easily be changed.
- Encourage greater customer focus, to facilitate a deeper understanding of what they want, and what is getting in the way of their satisfaction, and seek to act on this knowledge.

Creating Innovative Teams

Teams that successfully develop creative work have certain characteristics. The balance of these features has a notable impact on the team's capacity to generate creative ideas, and as a result, the make-up and design of teams is something that should be carefully configured.

Diversity is an important factor. Innovation is more likely when diverse people with diverse perspectives come together to solve a problem, and when individuals within the team feel safe enough to embrace their own diversity.

Diversity can, of course, bring with it challenges and disadvantages, particularly if trust and mutual respect is not developed. Diverse teams can get into conflict early on (the storming phase of team development) and if this is not dealt with effectively, it can cause the team to avoid healthy debate (and, as Ekvall noted, robust debate of ideas is a driver of innovation).

A classic tell-tale sign is when meetings are polite, but the real conversations take place in the corridor afterwards, with team members splitting into factions.

Team size is another factor that can help or hinder a team's creative capabilities. Teams of 3 or 4 members are easier to manage, conversation tends to be easier and team members are more relaxed, but there may not be the diversity of perspectives and talents present to create synergistic and creative dialogues.

Alternatively, when teams get bigger than 8 to 10 members, inhibitions start to become more apparent, especially for the more introverted team members, and dialogue that explores new and out-there ideas, or engages in constructive debate, becomes limited.

Beyond diversity, there are several other characteristics of successful creative teams. One factor that characterises a cohesive team rather than a set of individuals is the presence of a high degree of interdependency. People help each other out, even when their own work might take a lower priority for the moment.

To illustrate the benefit of interdependency, there is a game called 'broken squares' (see appendix) that can be a useful tool in team training exercises. Team members each hold several parts of a puzzle, and the team is asked to create a complete square in front of each team member by swapping pieces. Team members are not allowed to speak, nor can they request a specific piece from any other member.

The solution to the game is only possible when individuals are prepared to look at what others need, and give their pieces of the puzzle to others in the trusting assumption that others will help them out too. The game sometimes ends in stalemate when participants see it only in trading terms; I will only give you a piece if you reciprocate.

Such behaviour works if a direct swap is possible, but the game usually involves giving multiple pieces, so a reliance on one-to-one trading taking place does not lead to the solution.

Another direct parallel in terms of encouraging innovation is to get people, information and resources moving across organisational boundaries in a fluid way, based on a spirit of cooperation and a high trust environment. Solutions become possible when people are prepared to see the big picture, understand a problem in its entirety, and consider what they contribute to solving the problem, rather than focussing solely on how they can protect their own interests.

Trust takes time to build and, as the old Chinese saying goes, 'only a moment to lose'. Performance management systems which overly concentrate on individual results can hinder such fluid interdependent behaviour.

Before we look at improving teams, a word of caution: depending on their personality, people will react differently to team environments. For some, a degree of solitude is required to periodically recharge the batteries and to think more deeply about an issue, to reflect, make connections and find meaning.

Science fiction author Isaac Asimov (20) wrote, "The presence of others can only inhibit this (creative) process since creation is embarrassing. For every new good idea you have, there are ten thousand foolish ones, which you naturally do not care to display." So finding a balance between using team scenarios and giving people opportunities to work alone or spend time away from the team is necessary.

Here are some further considerations for creating and managing innovative teams:

1. Ensure that teams have members with a mix of skills and experience. Include those with little or no experience as well, and encourage the team to approach the project with an open mind. Indeed, those team members with no relevant background should be encouraged to contribute ideas from a novel perspective, as they may challenge some of the norms and assumptions of more experienced team members in a constructive way.
2. Continually work at building trust. Encourage team members to value each other. Spend time together getting to know each other. Teams are successful—and able to successfully harness their diversity—when they acknowledge and are respectful of the unique perspective, knowledge and expertise that each member brings to the table. This can sometimes require people to let go of their biases and prejudices.

3. In particular, recognise and accommodate the cultural differences within diverse teams. For example, some cultures place more emphasis on getting to know each other *before* conducting business (an Arabic saying advises, 'don't discuss business until the 3rd cup of coffee').

    Other cultures place the emphasis on learning about each other *through* working collaboratively on a task. Be mindful also of the unintended consequences of differing cultural interpretations. One company attempted to instil a sense of ownership and responsibility among its employees by encouraging them to act like a business owner.

    However, they failed to realise that in some cultures, acting like the owner entailed playing golf all day (21). Cultural hierarchies around seniority, age, and gender may make it harder for certain team members to speak up, particularly when challenging someone deemed to be more 'senior' in the hierarchy (e.g., a fellow team member who is older). Some of the techniques discussed in Chapter 5 can be used to overcome these hurdles.

4. Work at building early success by setting short term goals that stretch the team but are also achievable. A key characteristic of innovative teams is that they have clear shared goals and are committed to achieving them. All team members are committed to these goals, and share a level of excitement and motivation to achieve them. Look for ways for the team to celebrate their success in some memorable but appropriate way; the memory of a pleasant shared experience helps to build in-group solidarity and a team identity.

5. Team leaders must role model a perspective that sees mistakes or failures as a learning opportunity rather than a case for apportioning blame.

6. Agree ground rules for the team to adhere to. These ground rules should be agreed collectively by the team, rather than set by any particular team member, or by someone outside the team. One particular team that was established to solve a specific problem set the following rules: 'All ideas are good; everyone's voice needs to be heard; we treat every failure as a learning opportunity'.

7. Ensure that good facilitation skills are employed to harness the collective intelligence and to avoid group think. We discuss effective approaches to group facilitation in Chapter 7.

8. Be prepared to challenge organisational rules if they are getting in the way of new thinking. Be mindful, however, of the times when challenging these rules could be beneficial, and the times when it could be problematic, particularly for safety or legal reasons. Seek higher approval if the latter is the case!
A strategy of setting up 'skunk work' teams, as we discussed earlier, is designed to circumvent some of the organisational rules that might hinder the team's effectiveness.

A Model of Team Development

We must indeed all hang together, or, most assuredly, we shall all hang separately. – Benjamin Franklin to John Hancock at the signing of the Declaration of Independence, 4th July, 1776

Creating successful teams quickly has often been a major component of business success. And in an age of rapidly changing organisations, it has never been more critical. Yet team-building processes often get no further than a fun day out followed by a few drinks.

Alongside the considerations for creating and managing innovative teams described above, there are several stages to team development that teams must successfully navigate in order to build team effectiveness and continue to thrive.

Model of Team Development
Stages of Building a Team

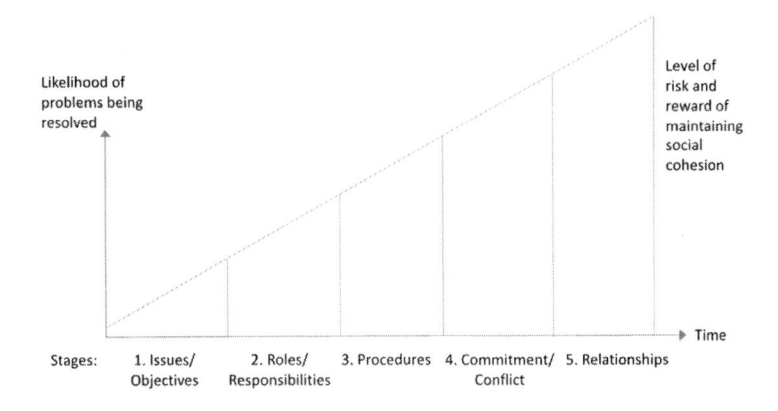

Likelihood of problems being resolved

Level of risk and reward of maintaining social cohesion

Stages: 1. Issues/Objectives 2. Roles/Responsibilities 3. Procedures 4. Commitment/Conflict 5. Relationships

Time

This model suggests that teams pass through five necessary stages in the process of effective team development:

1. Agree issues/objectives
2. Clarify roles and responsibilities
3. Develop efficient procedures

4. Work on team commitment and conflict
5. Improve personal relationships

1. Agree issues/objectives

The first stage is to clarify the goals of the team and the strategy to accomplish these goals. Team development should start here, with a discussion of the vision, mission and goals of the team. What do the team members want to accomplish? What is driving them? What threats and competitive pressures do they face that they must respond to? What are the incentives that will help to sustain effort to achieve the goals?

Trust and confidence in a team stem as much from *how* the team tackles such issues as it does from *what* issues it tackles. If goals and strategies are imposed without discussion, then enthusiasm and commitment are diminished. This mirrors the significance of freedom and autonomy on an individual level that we discussed earlier in this chapter.

Whilst teams may be set an overarching goal, the more freedom they can be given to determine how they will achieve that goal the better. Careful facilitation to engage and enthuse team members at the outset of the team's development is an important ingredient in building subsequent team resilience in the face of adversity.

In setting goals, a balance has to be struck to excite and motivate people with the big challenge but not demoralise them if it seems impossible. Incremental sub-goals and targets can be helpful.

An Olympic athlete believed that he had to run much faster than he had ever done before if he stood any chance of winning an Olympic medal in a few years' time. He and his coach predicted the time they thought would be necessary to win a medal and then broke down the necessary improvement into small incremental targets that were increased every three months. On the training track, the coach installed a red and green light half way, adjusted so that if the athlete was on schedule, a green light would come on and if he wasn't, a red one would illuminate instead. The athlete then knew he needed to run faster! At the subsequent Olympic Games, he won a medal.

2. Clarify roles and responsibilities

The second stage is to establish role responsibilities and relationships. Until there is clarity about who is doing what, the team rests on shaky foundations. The team therefore needs to work on a process of reviewing—together—who does what, how each activity relates to the team's shared goals and, most importantly, what is falling through the gaps between team members and thus not getting done.

This process results in clarity about role responsibilities and who relies upon whom for what, as well as the boundaries between roles to avoid the twin problems of 'turf wars' and 'organisation black holes'. Creating an understanding of who relies upon whom helps to establish a greater sense of responsibility to the team.

Throughout this process, the team roles should be adjusted to maximise the talent available within the team, as well as to minimise any gaps or weaknesses.

Furthermore, we must keep in mind that this clarity must also be combined with a level of flexibility. A rigidity and over-prescription of roles can prevent the fluidity that is required in 'team flex', as well as the agility that can be required in the process of championing new ideas.

3. Develop efficient procedures

Working together involves several generic processes that all teams must establish, as well as those processes that are unique to the team and its issues. Generic processes include decision-making, allocation of resources, and dealing with differences of viewpoint and absent members. There is a range of options for establishing these processes, from prescribed and fixed rules and procedures to more flexible approaches.

Teams might have strong leaders who decide how things are to be done, or the team might decide through more democratic means. There is not necessarily a right way, but what's crucial to success is the active agreement and recognition by team members that their mode of operating is right for them. In particular, teams need to find the right balance—which of course can change over time—between planning and action.

Teams can experience the 'risky shift' where they become risk averse and spend too much time planning every eventuality, or they can experience the reverse, becoming too risky and rushing too quickly into action and trusting in good fortune alone. Regardless of the final outcome, careful facilitation of these processes and discussions helps to achieve conscious agreement on a course of action, rather than condemning teams to blindly fall back on a default approach.

4. Work on team commitment and conflict

All teams have their own shared jokes and rituals, which help to create a sense of solidarity and exclusiveness. The balance between commitment to the team and commitment to the wider organisation is a tension that needs careful managing if unintended negative consequences are to be avoided.

For example, when team commitment is too high, other parts of the organisation, or teams in which members also play a part, are excluded or even denigrated, and unproductive conflict can arise. But when team commitment is too low, the team's effectiveness suffers considerably.

Creating teams in which members are sufficiently confident in each other requires working upon the dynamics of their interaction. Whilst this model of team development indicates that this is the fourth stage of the process, considerable progress should have been made during the first two stages, if the process has been effectively facilitated.

Building commitment starts by creating a sense of belonging for each team member. Do they have a part to play in the overall goal? Helping people make a connection between their place in the team and the successful achievement of the goal gives people a clarity of purpose and a voice.

In facilitating team meetings, the leader/chair/facilitator needs to recognise that confidence comes from being valued. Tough conversations may be required, but they will be more readily accepted if people feel valued. Employing a style that is 'tough on the issue but supportive of the person' can help avoid the defensiveness and lack of openness that commonly arises when people confuse feedback with personal criticism.

Dealing with conflict in teams in a way that leads to increased commitment can be difficult if the emotions surrounding conflict are not dealt with properly. Conflict can easily arise during the process of decision-making, and teams must clarify the process by which they make decisions. Approaches to team decision-making include:

- Consensus
  - This takes time to build but leads to the highest commitment.
- Majority voting
- Voting with a threshold that is higher than simple majority (e.g., 75% of votes)
  - This can be a valuable approach if the team must take account of minority views.
- Decision by the team leader, after consultation with the team
- Decision by the team leader, without consultation with the team
  - This can be valuable or necessary in situations where increased speed is required.
  - However such decisions are likely to have looser commitment from the team and a greater chance of being disputed if challenges arise.
  - Prior agreement about when it is appropriate to make decisions without consultation can lessen the possible resistance.

5. *Improve personal relationships*

Finally, with the basics in place, team development processes can focus on improving the personal relationships within the team. If the previous stages have been approached effectively, relationships will already be enhanced. Other activities that can help improve personal relationships include team members…

- working together on specific business issues for the team
- engaging in social events and fun activities, especially those in which they might be competing with other teams
- spending time together—professionally or socially—building a memory of shared experiences

We should not lose sight of the fact that every act of spending time together and working on the above processes and issues of team development increases team cohesiveness and effectiveness. In doing so, relationships are built and deepened, individuals become confident with each other, and there are less 'no go areas'—subjects that are ignored or avoided.

A successful team:

- Has a clear purpose
- Has authority to act
- Has the resources it needs
- Has the active support of its sponsor
- Knows exactly where it wants to go to
- Knows how it plans to get there, and when it will do so

A Checklist for Innovative Teams

The checklist below includes the features of teams that are innovative and synergistic. It can be used as a guideline in constructing teams, or as a 'health check' with existing teams, to identify ways that the team's operation or performance might be enhanced. The assessment can be conducted by an independent observer or as a survey of team members.

Characteristic	Assessment: Positive/ Negative	Action
Team members want to be part of the team, and are excited by the subject matter and the challenge presented.		
There are diverse skills and backgrounds in the team (not just age and ethnicity) that are well understood, accepted and respected by team members.		
Team members are willing to tolerate and work constructively with differences.		
Team members focus on what each can contribute, rather than clinging to narrowly defined job descriptions.		
Team events are carefully facilitated to ensure: 1. Maximum contribution from everyone 2. All ideas are captured 3. Crazy ideas are celebrated and viewed as possible routes to new practical ideas		

There are flexible boundaries between jobs that are negotiated (i.e., jobs in teams are adjusted to maximise the talent available and minimise any gaps or weaknesses).							
Processes are in place for capturing all ideas and working with them to produce practical results.							
There are recognition and rewards (however small) for innovation.							
The team is built around specific issues (e.g., problems to be solved, improvements to be made, costs to be reduced).							
Internal politics, such as game playing, are low level or, ideally, absent.							
There is a spirit of friendly rivalry coupled with a strong ethos of collaboration that builds team cohesion.							
There is a climate of lively debate and discussion, with no topics that cannot be discussed.							
Team members are mutually supportive and willing to subordinate their own needs to meet those of other team members.							

During the construction of the Heathrow Express train at London Heathrow airport, the tunnel collapsed, forcing delays and necessitating a radical rethink of the way the involved parties operated. They formed joint teams, that included team members from each contractor, to unite around a common goal of completing the tunnel, and they organised workshops and a 'suppliers club'.

This broke down traditional barriers and suspicions between contractors and got them to move beyond the adversarial blame-focussed approach to problem-solving. The results were impressive in terms of cost-savings and recovering lost time in the project.

Additional major initiatives that were undertaken, such as paying for training for subcontractors, contributed to shifting the attitudes of the workforce further. The lesson from this incident was that changing the way teams operate can significantly enhance their capacity to solve problems.

Rewarding Innovation

Consider this question: What is actually favoured more within the organisation—undertaking actions that might bring about novel solutions, or simply maintaining the status quo? The answer to this question may well lean towards the novel solutions in theory, but to uncover the true answer, it is necessary to look at the system of explicit and implicit rewards to see which way judgement goes.

How are people promoted? Who gets the bonuses? Or the favourable ear of the boss? Invariably the downside of getting things wrong has harsher consequences than the possible upside that the new approach might be beneficial, so there is a strong pull to value actions that are safe or maintain the status quo.

Across the organisation, therefore, it is valuable to consider what behaviours are rewarded—explicitly and implicitly. Are these the behaviours that you want to encourage? Are these the behaviours that support all the different approaches to creativity and innovation that we have discussed?

Mindful of the behaviours that you *do* want to encourage and the approaches to business and innovation that you *do* want to see flourish, what are the changes to this embedded reward system that you need to change?

Prizes have been a recurring theme in innovation from the development of the railways in 19th century Britain to Alcock and Brown's first non-stop transatlantic flight in 1919, which won them £10,000 from the *Daily Mail* newspaper. However, thanks to a number of studies, it has become evident that money is not necessarily the best motivator of performance and innovation, particularly at an organisational level, nor are other extrinsic (external) motivators such as the promise of promotion.

Instead, as we have discussed, it is *intrinsic* motivation that is key to performance, perseverance, and achievement in creative endeavours. Author, and former vice-presidential speechwriter, Dan Pink (22) talks of 3 key ingredients in fostering intrinsic motivation: autonomy (having the freedom to act), mastery (having the ability to learn and become increasingly proficient at what you do) and purpose (having a compelling vision that you feel is worth striving for).

This calls into question the nature of reward, and how performance, effort, and creative achievement should be recognised and rewarded appropriately in order to encourage continued efforts. There are a number of issues to take into account when considering the nature of these rewards within organisations.

Where there is a reward system which is overly based on successful performance and successful outcomes, it causes people to focus on the easiest way to reap the rewards and avoid the punishments (usually what has worked in the past), rather than focussing on the activities and efforts required for innovation.

A climate of fear of getting it wrong will invariably lead to playing safe, entirely eliminating any opportunity for creativity and innovation. Instead, reward and recognition should be bestowed upon the creative efforts, regardless of the outcome—which is often out of our control.

Praise and recognition for effort and persistence, not just success, are important. Since failure is a fundamental part of innovation, it should not be viewed as a disqualifier for praise, recognition, and reward. Instead, it should be acknowledged that failure is a powerful teacher and that future attempts will be all the better for the learning, provided the learning actually occurs.

Performance management evaluations that recognise failure in innovation as a result of trying new ideas are to be encouraged. Indeed, Thomas Watson, former chairman and CEO of IBM once said, "If you want to increase your success rate, double your failure rate."

One day, an employee made a big mistake, which cost his company significant money. He decided to resign before he was fired but when he handed in his resignation, his boss tore it up, saying, "I have just spent a lot of money on your training, I don't want to lose you now."

Note that it can be very useful to maintain suitable records of all ideas and failed attempts. It may be that their time has not yet come, but may do so in the future. A study of previous patents highlights the crucial role of timing to an idea being seen as useful.

Frank Hornby patented Meccano in 1901, but it was some years before the construction toy became ubiquitous, its popularity spawning a range of other construction toys, to eventually be surpassed by Lego.

In the same year, a Mr Blunt patented his 'improved flying machine', which was powered by steam-driven wings. At the time, given the level of knowledge and interest in aeronautics, it would have been difficult to guess which patent would be the most successful.

Computing and IT company Hewlett Packard (HP) used to throw a party whenever a project that they had launched finally came to its demise and was cancelled. The company wanted to recognise the efforts that had gone into the project and signal that these efforts and attempts were valuable and worthy of recognition and reward just as much as a successful product.

Furthermore, HP also had an award for the person who had succeeded *despite* the organisation (recognising that organisations often unwittingly hinder or thwart creative efforts). In the early development of the oscilloscope, an HP manager backed an employee who believed they could develop a marketable product, despite the decision of senior management to shut down the project. The employee succeeded, and HP's senior management forgave the employee and their manager for their deliberate flouting of the directive to shut down the project.

Cultivating a sense of accomplishment throughout the process of innovation is powerful in encouraging ongoing effort. Even though the final destination (of success or otherwise) may be a long way off, marking certain milestones with a celebration or some other form of recognition, can demonstrate the value an organisation places on the process of innovation and on the project itself, and give staff a powerful boost in their sense of value and purpose, and their level of motivation.

There is an important distinction between the use of money as an incentive (which is generally not effective in motivating creative endeavours) and a fair reward for the effort and, at some point, the achievement of successful innovation. It is important to ensure that employees feel valued in their contributions and their endeavours, and providing recognition that is financial in nature, such as a share in the profits of the innovation in question, can be powerful.

On the subject of finances, pay structures such as broad banding, that allow flexibility to reward innovation regardless of other characteristics that determine the pay scale (such as managerial level), can be effective. Narrow grade bands where researchers are often at their ceiling are a source of niggle. Pay that recognises skill, knowledge and experience—and recognises that skilled researchers may be paid more than their managers—is a powerful symbol. Furthermore, with regard to advancement, it may be worth considering how to carefully remove those who block creativity and promote those who champion it.

Resource Allocation and Control

Tales of great innovations also abound in tales of failure along the way, and it requires great faith on the part of organisational management, together with a willingness to see resources bearing little fruit, to support innovative projects until breakthrough occurs. In the midst of such disappointing failures, of course, there is no guarantee that a breakthrough will occur.

All too often, we call for a business plan to evaluate a proposal in its need for resources. Yet by the very unpredictable nature of innovative projects, this common approach will not be effective. Instead, organisations can use a variety of approaches for projects with a lower potential of success, including bidding for project finance from a specific innovation fund. Again, there is no expectation of a quick return and organisations must accept Henry Ford's maxim that "failure is a perfectly acceptable option".

Evaluating a bid for resources when there is no definite chance of success is a risk that many an executive is not prepared to take. Yet resources (time, money, personnel, expertise) are all crucial elements in the mix of innovation.

Possible Criteria for Evaluating a Bid for Resources for an Innovative Project

In the list below, we identify a number of points that might be useful when evaluating a proposal for resources.

Many previous innovations have broken all the rules. So use this checklist only as a guide, not a definitive go/no go decision maker. Ultimately, judgement and courage will come into the decision.

1. What is the track record of the proponent? Have they shown creative ideas in the past?
2. What resources are required? Will the idea of 'seed' money for a defined period prove a useful concept?
3. Does the development fit with the core competence of the group developing it?
4. What is the likely impact of the innovation if successful?
5. What is the opportunity cost of expending these resources? What is foregone?
6. Are there any lead indicators that the project might be successful?
7. What assumptions are being made? (For example, viability of technology, market size, abilities of resource providers, willingness of key stakeholders, etc.)
8. Are there low risk routes to prove the concept viable?
9. How easy is it to make multiple refinements and improvements from the first concept?
10. How rapidly can we learn from experiments and so improve the possibility of success?
11. When is the go/no go decision? When do you pull the plug? (Beware the danger of requests for 'just a little bit more time, money, etc.')

Architecture and Building Design

Can buildings influence creativity? According to research by Groves and Knight (23), building design can facilitate stimulation, reflection, inquiry, collaboration, and play. Whilst creative ideas can come to mind at any time, and in any place, there are approaches to designing spaces that can make them better equipped to support and encourage the creative process.

Researcher John Kao (24) described such spaces as, "safe, casual, liberating. Not so small as to be limiting, not so big as to kill intimacy. Creature comfortable, stimulating, free of distractions and intrusions. Not too open, not too closed; sometimes schedule bound, sometimes not." As a result, numerous companies have experimented with a range of approaches to office design.

Canon (electronics) has meditation rooms; Shiseido (skin care) has walking paths and trails on its campus (25); Orange (mobile phones) has a room with bean bags, games consoles and a wide selection of music. With its Googleplex, Google made a deliberate attempt to create a fun and stimulating environment. The Nerdery, a software house, has created multiple spaces that are tailored to suit different times and different moods.

When Cavendish Partners, an outplacement and counselling firm based in London, established their offices, the partners decided that they should choose furniture that would help create a safe environment. The chairs were solid and comfortable, with large and supportive arms; this had the effect of helping the clients to feel more of a sense of ease and safety.

These approaches aim to provide employees with an environment that will stimulate their senses appropriately (via calming colours on the wall or access to natural light and nature, for example), to help cater to their personal conditions for creativity, as well as to their sense of freedom and autonomy to do the work in the way that suits them best.

It is a useful question to ask of employees: What is the ideal work environment for you, and in what circumstances are certain factors important (recognising that this may vary depending on the activity)?

Another key factor is to provide an opportunity for interaction, particularly across teams. Steve Jobs famously designed a Pixar office building with the one set of bathrooms in the central atrium of the building, so that everyone would have to come to the middle of the building several times a day, and in doing so, were likely to run into people from other departments and have impromptu discussions that could lead to interesting ideas and insights (a second set of bathrooms was subsequently added to the design!).

However, other Pixar buildings weren't necessarily so successful. Commenting on one particular building, Ed Catmull said, "It's a mistake. It doesn't create the kind of interactions we need to create. We should have made the hallways wider. We should have made the café bigger, to draw more people. We should have put the offices around the edges to create more shared space in the centre. So it wasn't like there was one mistake, there was really a lot of mistakes, along of course with the bigger mistake that we didn't see most of the mistakes until it was too late" (26).

The office of UK airline British Airways, at Waterside near Heathrow airport (and designed by acclaimed Norwegian architect Niels Torp), has a 'street' running through it, as a mechanism for encouraging interaction between people from all areas of the business.

However, it is important to recognise that the design of office buildings is not a driver of innovative cultures on its own; such design must be accompanied by an organisational culture that supports and encourages the use of the various features of building design, and values the potential outcomes accordingly.

If an office incorporates outdoor spaces designed to facilitate reflection and stimulation in nature, it must not then establish cultural barriers that prevent employees from actually using these spaces (e.g., disapproval based on an underlying belief that 'real' work only takes place at one's desk). Without the supportive culture in place, simply painting the walls a different colour, and providing toys and easy chairs does not result in creative ideas.

One further caveat—people who are used to more traditional workspaces may well react negatively against such changes. As such, they must be allowed to use the office in whichever way best suits their work practices. Some changes may work for some people and not others, or in some situations and not others; a process of inclusive consultation with staff as part of the building design can help to tease out any potential hurdles here.

There are also various elements of architecture design that can explicitly hinder innovative efforts within organisations, such as a strong prevalence of individual offices and the use of tall tower blocks with many different floors. These conditions prevent people from interacting easily, particularly across business divisions and teams.

Hewlett Packard's offices in Bristol (UK) dealt with this by only having two floors and being predominantly open plan, with easily movable furniture and walls. This flexibility allowed teams to be physically located and relocated (and furniture and walls moved accordingly) as priorities and projects changed.

For some people, open plan offices have facilitated creativity by increasing communication and interaction. But for others who might require some solitude or uninterrupted periods of work, open plan offices can be highly irritating and disruptive, and ultimately detrimental to their capacity to do creative work. Suffice to say, the space needs to fit both individual personalities and preferences as well as the specific problem being tackled.

Helpers and Hinderers for an Innovative Culture

Based on the various factors we have discussed throughout this chapter, the table below can be used as a diagnostic tool to assess the state of an innovative culture within any organisation.

Helpers	Hinderers	Assessment of your position: Positive/ Negative	Possible action
Meaningful purpose and mission.	Lack of relevance of mission.		
Easy to change team goals.	Goals set for a long period regardless of utility.		
Evaluating performance on activity and results.	Evaluating performance on attendance and results.		
Empowerment of individuals and teams.	Strong control of activity.		
Mistakes and failures treated as learning opportunities.	Culture of fear about getting things wrong. Culture of blame for the 'culprit'.		
Personal freedom in job design to pursue ideas that are not strictly part of the job. Roles include an expectation of looking for improvements and new ways of doing things.	Overly prescriptive job descriptions and strong boundaries dictating whether a task is allowed or not. Specialisation, compartmentalisation and standardisation.		
Diverse skills and experience in work teams.	Recruitment follows profile of existing team members.		
Flat structures with fast decision-making.	Long hierarchies and bureaucratic decision-making.		

Free flow of information.	Filtering to give people only 'what they need to know'.		
Style of management is one of coaching and facilitating.	Style of management is one of control and micromanagement.		
Collaboration across organisational boundaries.	Silo mentality. Culture of destructive criticism and competition.		
Belief that there is freedom for the individual as well as teamwork in all areas of the business.	Belief that the 'lone wolf finds the solution', or that innovation is the sole province of the 'new product division'.		
High challenge, high autonomy.	Low challenge, high routine and repetition.		
Staff are engaged and satisfied with their work. Thinking time and working time are congruent.	Staff are disengaged and thinking about other things during work time. Rarely think about work outside working time.		
Willingness to create temporary teams (natural work teams) to examine and solve problems.	Only operating within existing hierarchies.		

Chapter Summary and Conclusions

In this chapter, we have focussed on the organisational conditions that encourage innovation. In particular, we have explored:

- Leadership for creative cultures and the approaches that leaders can adopt that both help or hinder creativity.
- Job and organisation design with facets of roles and structures that facilitate greater flexibility.
- Innovative teams and approaches to building them successfully.
- Resources and rewards, and the need to recognise efforts as well as outcomes.
- How architecture and the layout of buildings can enhance opportunities for creativity to flourish.

Chapter 7
Designing Problem-Solving Workshops and Other Creative Events

Introduction

As we have discussed throughout this book, we believe that the process of creativity is no accident, even when it has appeared in a flash of inspiration. More often than not creativity is the result of a process that can be enhanced. In this chapter we look at the facilitation of events that seek to stimulate creativity and innovation.

In particular, we focus on events where people come together and seek to solve problems and generate new ideas. As we have discussed earlier, coming up with novel ideas and creative solutions can often be an individual activity, and a quite solitary one.

However, group and team approaches to creativity can also be extremely productive. Whilst we should not assume that groups will always be better, it is worth recognising that groups can sometimes, if planned and facilitated well, act synergistically and come up with better ideas than any one individual.

Balancing the tension between giving people space to be creative alone, and gathering people in groups to harness their synergistic power, can be an interesting challenge for teams and organisations.

This chapter focusses on the use of more structured approaches among groups of people and sets out processes to help new ideas to emerge when groups are focussed on solving problems.

Setting Up a Group Process

Designing workshops and events to encourage creative thinking requires a number of factors, and a consideration of the many elements of the creative process that we have discussed throughout this book.

The comments of Avril Loveless, of the School of Education at the University of Brighton (1), on the processes and factors of a creative environment within schools, may serve as a useful recap here. Whilst they are focussed on education, these considerations are pertinent to group approaches in any situation. She highlights five characteristics of a creative process: "using imagination, a fashioning process (i.e., producing something), pursuing purpose, being original and judging value." And she includes the following factors in her prescription for a creative environment:

- Awareness of the ways in which creativity is related to knowledge across the curriculum [or organisation].
- Opportunities for exploration and play with material information and ideas.
- Opportunities to take risks and make mistakes in a non-threatening atmosphere.
- Opportunities for reflection, resourcefulness and resilience.
- Flexibility in time and space for the different stages of creative activity.
- Sensitivity to the values of education [or the organisation] which underpin individual and local interest, commitment, potential and quality of life.

(We have added 'the organisation' in parentheses to illustrate these points' application to such an environment.)

Prior to any event or workshop, consider whether the process might be enhanced by individuals doing their own research and reading first, in order to come to the event already primed with ideas and appropriate knowledge.

There is a balance to be struck in such prior priming. We want to give participants the opportunity to develop ideas based on information from multiple sources, and to allow such information to incubate before the event or workshop.

However, we do not want participants to become so entrenched and invested in the ideas they bring to the event that they seek to defend these views rather than participating in a more open discussion, where ideas are evolving and views can change.

In Chapter 6, we identified the critical role management behaviour plays in creating the culture and environment for ideas to emerge. No less important is the organisational commitment to pursue creative activities and to invest time and money in the endeavour.

Chapter 6 also discussed a number of factors and steps in the process of establishing and managing ongoing teams for creativity, innovation, and problem-solving. In this chapter, we revisit some of these points, but look deeper into the process of running team events and workshops for the purposes of creative problem-solving.

Participants

In Chapter 6, we discussed the optimum number of people in a team, and the need to strike a balance between the quest to gather as many ideas from as many sources with the practicalities of effectively managing a group. The question we now turn to is the ideal composition of team characteristics to achieve creative outcomes.

Beyond a consideration of diversity in the group, the choice of participants in any group process should be based on a number of characteristics.

Knowledge of the subject. Some knowledge or experience of the issue or problem that is the focus of the workshop is key. As we have discussed, knowledge, expertise, and experience bring with them the raw materials to explore creative ideas. Of course, anchoring can pose a risk, and those with no knowledge but a fresh eye can be very helpful in developing new ideas. Ideally, aim for a balance in the group of those with domain expertise and those without.

In a German car plant, a consultant was asked to advise on improvements. He met the board and asked them to place their shoes on the boardroom table. The consultant inspected the shoes and declared, "These shoes have not walked on the factory floor, only the director's corridor. Once there is evidence that you understand your business from the bottom up, then I will work with you."

Desire and motivation to develop new ideas. Where is the passion? Members of the group must be passionate about solving the problem and motivated to find solutions. Volunteers are likely to be more intrinsically motivated than those who are told to participate.

Commitment to the process. It is important that participants have a degree of commitment to the process, to the team, and to the success of the outcome. 'Grandstanding', when people engage in team activities solely to show off or boost their own standing, or merely observing with no contribution to the process or outcome, can be undermining. Clarity about commitment up front is important to establish, as committed people are more likely to overcome the inevitable challenges of implementation.

Cross-functional representation. It is particularly important to have team members from different areas and functions of the business when the problem to be solved involves and affects many different areas. In Chapter 6, we discussed the creation of 'natural work teams' as an approach to forming cross-functional teams.

Ownership of the necessary resources to bring about change and to implement any solutions. Who has ownership of the necessary resources to bring about change and to implement any solutions? Ensure that the team is empowered to implement change, or bring solutions to those who can then implement them.

Ground Rules

When using these tools and techniques in group situations, it is important to establish ground rules to keep the conversation and interaction productive.

Typical ground rules include:

- Generous listening
- Plain speaking
- Constructive challenging
- Build on others' contributions
- No subjects are taboo
- No idea is 'stupid'; every idea has value
- Separate the process of idea generation from the process of evaluation
- 'Chatham house rules' (what is said in the meeting can be quoted outside but not attributed to any person; as opposed to 'confidentiality' where nothing is taken outside the room)
- Keep to time

Group Dynamics

Chapter 6 highlights certain group dynamics that can arise in team situations. When groups come together for problem-solving workshops or events, there are two dynamics that we must watch out for in particular. The first is 'the risky shift', which we described earlier, and the second is 'follow my leader': when there are senior figures or key opinion formers in a group, they are more likely to have greater influence on the collective opinions and decisions of the group.

This tends to neutralise the synergistic benefits of groups, as other team members become less willing to voice their own ideas or perspectives, particularly if they are at odds with those of the senior figures, and potentially give up trying to think for themselves altogether, simply deferring to the contributions of the senior figures.

Cognitive Differences

Those familiar with the Myers Briggs Type Indicator[16] and other personality inventories will recognise the different preferences that occur between different personality types. For example, typical introverts prefer to think before speaking, whereas extraverts are frequently comfortable to effectively 'think aloud'. As a result, group brainstorming tends to work well for extraverts, but introverts may prefer an approach like brainwriting (see Chapter 5) that allows them time to think and facilitates equal contribution from all parties.

Another difference that can pose challenges in team situations is between those who need to understand the big picture before diving into the detail of an issue, and those who need to get the detail right before they can navigate the big picture. This difference can become more pronounced when there are errors in the information being examined.

For the more detail-conscious, where 'the devil is in the detail', there seems little point in pressing on until the detail is right. On the other hand, the big picture person gets frustrated at getting bogged down in discussions of detail when they want to clarify the overall situation first.

When it comes to making decisions between competing ideas, there will be those who are more swayed by logic (for example, convinced by numbers and

---

[16] Resources to work with the Myers Briggs Type Indicator can be found at www.myersbriggs.org

statistics) whereas others will be more affected by their values and instincts of what is right (for example, convinced by their gut instinct and what 'feels right').

Finally, there are those who prefer to have a plan and take a more structured approach to moving ideas forward into implementation, whereas others prefer a more organic and emergent approach that is determined by what they discover during the process.

Accordingly, we must remember that there is no specific profile or preference that is required for creativity—anyone with any preference is capable of generating creative ideas. As a result, all these approaches must be respected and valued in the creative process.

"Mad hair. Perhaps a lab coat. Or questionable knitwear. Few inventors fit this stereotype, just as few inventions are flights of fancy worthy of Heath Robinson. Not that there's anything wrong with an elaborate contraption bristling with cogs, pulleys and water clocks, that boils a kettle while simultaneously tipping its owner out of bed and straight into a natty vest/corduroys combo. But often the best inventions come about because an ordinary person has a problem that needs solving" (2).

The challenge, therefore, is in acknowledging and accommodating the different preferences in a team situation. Workshops and groups tasked with coming up with ideas benefit from having a mix of styles, but facilitators must also recognise the difficulties that can arise. The aim is to ensure a balanced contribution from all team members, and the challenge is to keep these tensions between the different preferences creative rather than destructive.

A useful approach is to surface the differences and the resulting tensions in an unemotional way, so they become explicit and normalised. This then gives participants licence to embrace these differences, and expand the range of possibilities as a result.

Another area in which differences between team members' cognitive styles can create tension is in the approach to the learning process. Learning is an intrinsic part of the creative journey and individual differences in learning styles (see Chapter 3) will result in different preferences in how to approach the unknown. There will be some people who gravitate towards experiential learning and will be more tolerant of failure during the problem-solving process, as this is a powerful learning experience. Others will prefer a more theoretical approach and want to study what has happened in the past so that any experiment is conducted with the full knowledge of prior experience and the learnings from prior mistakes (3).

To help accommodate the different thinking styles, it can sometimes be useful to assign people specific roles that explicitly call on the different thinking styles (4). You might have an 'idea generator', whose sole task is to come up with lots of ideas, whilst a 'researcher' might look for all prior knowledge and evidence that might be useful in the problem-solving process. A 'completer finisher' might then play a vital role in attempting to bring together the loose ends.

The challenge for the facilitator (who may also be the group's manager) is to ensure that their own style doesn't get in the way of others speaking up and working on new ideas. Facilitators who prefer logical decisions and specific plans may find themselves being critical of those who generate ideas that might initially seem impractical, for example.

There is, of course, a time for more critical evaluation, but in the first flush of enthusiasm for a new idea, at a time when the idea is at its most fragile and when the speaker has risked ridicule to express it, what is needed is support, validation of the courage to speak out and encouragement to pursue it further, regardless of cognitive preference.

The facilitator's task is to keep the conversation open and progressing without early closure, despite the differences in team members, and even if frustration starts to build from navigating the differences, or from the challenge of generating more ideas. Motivation and morale must be maintained to push through when frustration might cause some people to give up contributing.

When there are signs of dissatisfaction or emotional withdrawal, a facilitator will need to point out (unemotionally) these signs so that the group can diagnose what is happening. The group can then decide whether to pause and take a break or push on.

When conflict arises, the aim of good facilitation is to ensure the conflict is not buried but is dealt with as unemotionally as possible. One approach to mediating conflict follows this process:

- Allow each person to clearly state their view whilst ensuring that everyone else listens.
  - This is designed to reduce the emotion that might have built up as conflict developed.
  - In conflict situations, high levels of emotion tend to inhibit the capacity to listen to other people's perspectives, as everyone is too busy thinking of their next argument, so explicitly asking everyone to listen is important.
- Allow each participant to talk (without interruption) about the impact that others' views and suggestions have on them, and the potential solutions they seek.
- Encourage discussion about the causes of the differences.
- Once causes are understood and jointly agreed, lead the search for solutions that address these causes and meet the needs of all participants.

A Generic Problem-Solving Process

There are many models and processes designed to create a structured approach to problem-solving. There is an inherent paradox in seeking to establish a process for finding creative ideas, given that creative ideas and insights can be so unpredictable in their genesis. Any process, therefore, will need to be alive to the fact that novel ideas that might spring from anywhere and at any point in the process, and might therefore undermine or derail the process itself.

The aim is to hold on to the steps of the process very lightly, and be prepared to adapt and follow creative inspiration wherever it leads. Any process is simply an aid to finding solutions, not an end in itself.

Approaches to group problem-solving often include the following generic process.

(Note: The tools mentioned are described in Chapter 5.)

Step 1: Name the Problem

Discuss the problem to be solved in order to build a shared understanding within the group. What are the symptoms of the problem? What are its dimensions and dynamics? Be careful how you name the problem and remember that how you name and frame the problem will affect the lens through which you envisage the possible solutions (remember how the difference in framing a problem as 'too many cars' versus 'too few parking spaces' can lead to different solutions to the problem).

It can be useful to remember Occam's razor[17], which seeks to avoid unnecessary complexity in hypotheses of problem and solution.

Effectively naming and exploring the problem involves firstly identifying the broad theme that encompasses the problem. It may be helpful to capture this in an affinity diagram, using the themes that arise from a brainstorm of problem areas. To narrow the problem down further, ask: Is it something we can influence? If not, what are the elements of the problem that we can influence? Can we make progress in a reasonable time? Can we collect data about it? Do we really want to solve it?

The next step is to craft a precise problem statement. Ensure that the group have a mutual understanding of this problem statement and a shared desire to solve it. A final step is to check whether it is possible to solve this problem given the time and resources available to tackle it.

Step 2: Probe to Identify What Might Be the Underlying Causes of the Problem

Now that you have clarified the precise problem to solve, start to examine it deeper to discern the underlying causes of the problem. Identify the chain of cause and effect that leads to the problem. Tools that will help with this process include a fishbone diagram, root cause analysis or the '5 whys' process.

Process mapping or flow charting might also be useful if appropriate, and utilising the 40 TRIZ principles can be helpful. Remind participants that there are no such things as stupid questions in this process.

Some useful prompts and questions are:

- 'Describe the problem in detail, step by step.'
- 'Imagine I know nothing and explain the problem to me.'
- 'What are the signs of this problem?'
- 'Under what conditions does this problem occur?'

---

[17] William of Occam was born circa 1280 and is known for his dictum, 'entia non sunt multiplicanda praeter necessitatem', which is interpreted today as 'things should be kept as simple and uncluttered as possible'.

- 'Who does this problem affect?'
- 'What are the consequences and particularly the unintended consequences?'
- 'What is stopping us from solving this problem?'

It is easy at this stage to make false assumptions about who or what is the cause of the problem. As we discussed in Chapter 5, too often the 'presenting problem' is not the real underlying problem, and the latter only becomes apparent with deep and careful analysis.

One of the authors spent his early career in human resources. One day he interviewed a woman with a poor sickness record with the intention of disciplinary action as there were no obvious signs of illness. His assistant pointed out that there was a pattern of a couple of days off each month and wondered whether it was related to her menstrual cycle. On probing the causes of illness (she reported feeling generally unwell), it was decided to send her for a scan, which revealed cervical cancer.

Step 3: Identify the Goal of Your Problem-Solving

What does success look like? What is the ideal solution? Is it a total and complete solution? Or does it generate a range of options? Or a partial improvement? Identify the specific goal of this problem-solving process, and be clear about what constitutes success. It can also be helpful to identify a fall-back goal if the ideal solution is not possible. Considering the next best solution can be very useful, as it might actually be good enough, particularly if you are getting bogged down in the challenges of delivering the ideal solution.

Don't lose sight of the problem you are solving, and the fact that there may be multiple solutions or measures of success, because you are so tied to the solution you have decided to aim for. Beware of missing the *true* measure of success: "the operation was successful, but the patient died" is not the desired outcome at a hospital, even if the problem was simply to perform a specific operation.

Step 4: Identify Possible Solutions

Use idea generation techniques, such as a particular approach to brainstorming, to generate ideas and possible solutions to the problem. Chapter 5 contains numerous techniques that can be helpful.

Remember the principles behind generating ideas:

- Experimentation is necessary and you should aim to generate as many ideas as possible. Aim for quality rather than obsessing about quality.
- A lack of censorship or judgement is important. All ideas should be welcome and recorded.
- Embrace the crazy, unusual, or seemingly impossible ideas. You are aiming to bring about new connections, new perspectives, new insights, and this requires wading into uncharted territory.
- Build on each other's ideas, bounce off each other, and find a sense of playfulness in the process.

Encourage broad, blue-sky, what-if thinking:

- 'What if money were no object?'
- 'What if you owned the company?'
- 'What would customers/suppliers/competitors suggest?'
- 'What would happen if we did the opposite of our current approach?'

Remember that creative nuggets can come from anywhere. It may be a chance remark, a joke, an ill-formed idea, but these may be the starting point for a creative solution, so ignore nothing at this stage.

During this phase of generating potential ideas to solve the problem, you might consider how you can throw the net for ideas as wide as possible. Supplement the ideas generated within the group with ideas drawn from other sources—competitors, literature, case studies, prior experiences in other situations.

It may be useful to employ an idea management system, which provides a software solution for capturing, storing, and managing a large number of ideas from a wide variety of sources. These systems allow a much wider group to participate in a creative process, and can allow an organisation to combine the generation and capture of ideas across the entire organisation with more focussed idea generation and problem-solving of dedicated interactive groups.

Step 5: Evaluate Solutions

Decide on the approach you will use to evaluate all the ideas and solutions you have generated…and then evaluate them! Chapter 5 discusses a number of approaches to idea evaluation, from team voting to impact analysis.

Step 6: Create an Action Plan

Once you have decided on a solution, you must clarify an action plan to pursue and implement this idea. The plan should be expressed using the SMART goal framework (discussed in Chapter 4).

At this point, accountabilities should be defined, as well as clarifying who has authority and discretion to take certain decisions or actions, particularly in regard to financial details and expenditure. The team should have clarity on who is driving the implementation process, and ensure that they have all the necessary authorities and resources to make it happen.

Step 7: Troubleshoot the Plan

Aim to anticipate any obstacles. A helpful tool to use might be the force field. Make sure that the group wrap up the process feeling positive about their work and the solutions that have emerged. They will be important ambassadors for future problem-solving groups, as well as for the successful implementation and long-term adoption of the solution the team has identified.

Step 8: Monitor and Evaluate the Plan

Establish a monitoring mechanism to track progress towards successful implementation of the solution. By monitoring progress, you can ensure that any corrective actions that are required for success can be taken.

A simple technique to evaluate the plan is to identify 'alpha, delta'—what has gone well that should be reinforced in the future, and what could be done better next time. Note that the focus here is not necessarily on what has gone 'badly', but rather aims to identify future opportunities using positive solution-focussed language.

Ensure those who monitor and review the implementation of the solution are not concerned with merely preserving the status quo. Give feedback to those that might form future problem-solving teams in the quest for continuous improvement in the process, as well as improved skills and teamwork.

A variation of this problem-solving process is designated by the acronym SCORE, whose stages are to identify the Symptoms of the problem, identify the Causes, clarify the desired Outcomes, identify options to Resolve the problem, and then Execute any actions.

A Note on Idea Management Systems

Historically, many large companies have relied upon suggestion box systems to gather ideas from their employees. But these systems often suffered from a number of common shortcomings. Because they weren't usually focussed on specific business goals, suggestion box systems tended to attract a small volume of low-quality ideas.

Once an employee submitted an idea, he or she usually never learned what became of it. As a result, employees often became cynical and would no longer contribute their ideas to the programme. Paper-based suggestion box systems also made it hard to ensure that all ideas were evaluated on a timely basis and in a consistent manner.

In contrast, idea management software tools are designed to help organisations focus their employees on specific business issues, and are customisable and campaign-focussed. This tends to result in a larger quantity of quality ideas. Also, because today's idea management systems are powered by databases, it is much easier to set up and manage a closed-loop evaluation process, which automatically reminds evaluators of upcoming deadlines and unevaluated ideas.

Idea management tools also focus employees' creative efforts around specific organisational goals and objectives. They encourage employees to capture all of their ideas, and collect ideas from all areas of the organisation. By placing ideas in a shared repository, idea management systems promote greater transparency.

Finally, they help companies to increase their speed to market and to share best practice.[18]

Other Approaches to Problem-Solving

A number of other models have been proposed as approaches to problem-solving. They tend to share underlying similarities with the generic approach described above, but they are worth examining as their unique elements may resonate more effectively in certain organisations and teams.

The Osborn Parnes Creative Problem-Solving Model (5)

After many years researching and teaching creativity, Alex Osborn, founder of the Creative Education Foundation (and the man credited with codifying the process of brainstorming), and Sidney Parnes, a famed creativity researcher and academic, formulated their own model for a creative problem-solving process.

Stages

1. Mess finding
   - Determining where the problem lies and establishing the objective of the process
2. Fact finding
   - Exploring the facts and details of the issue or problem to be tackled
3. Problem finding
   - Clarifying the exact problem to be solved
4. Idea finding
   - Generating ideas and possible solutions to the problem
5. Solution finding
   - Evaluating the ideas

---

[18] At time of writing, examples of idea management systems include Imaginitik and Invention Machine Goldfire.

6. Acceptance finding
   - Establishing a plan and the practical aspects of implementation

Mess Finding/Objective Finding

Use this checklist of questions prepared by Parnes to prod your thinking:

- What would you like to get out of life?
- What are your unfulfilled goals?
- What would you like to accomplish or achieve?
- What would you like to have?
- What would you like to do?
- What would you like to do better?
- What would you like to happen?
- In what ways are you inefficient?
- What would you like to organise in a better way?
- What ideas would you like to get going?
- What relationship would you like to improve?
- What would you like to get others to do?
- What takes too long?
- What is wasted?
- What barriers or bottlenecks exist?
- What do you wish you had more time for?
- What do you wish you had more money for?
- What makes you angry, tense or anxious?
- What do you complain about?

Fact Finding

Use Who, What, When, Where, Why and How questions to examine the facts and details of the situation.

- Who is or should be involved?
- What is or is not happening?
- When does or should this happen?
- Where does or doesn't this occur?
- Why does or doesn't it happen?

- How does or doesn't it occur?
- …and so on

Problem Finding

List alternative definitions of the problem. Earlier in the book we discussed the principle of creative problem-solving whereby the definition of a problem will determine the nature of the solutions. We suggested that it could be useful to look at the problem through different lenses: bodies of knowledge, language and cultural groups, stakeholders, even Martians!

In this step, asking some further questions from these different perspectives might help uncover a deeper understanding of the problem. From each perspective that you adopt (e.g., as a Martian), ask the following questions:

- What is the real problem?
- What is the main objective?
- What do you really want to accomplish?
- Why do you want to do this?

Idea Finding

This is the divergent-thinking, brainstorming stage. This is where a variety of idea generation techniques can be used (see Chapter 5). Ideas should be freely proposed—without criticism or evaluation—for each of the problem definitions identified in the previous stage.

Solution Finding

There are three steps:

1. Clarify the criteria for evaluating ideas (see the selection grid and impact analysis techniques in Chapter 5)
2. Evaluate all the ideas generated using the chosen criteria
3. Select one or more of the best ideas according to the evaluation

Criteria might include:

- Will it work?
- Is it legal?
- Are the materials and technology available?
- Are the costs acceptable?
- Will the public accept it?
- Will higher-level administrators accept it?

Acceptance Finding

At this stage, plans are made to move the ideas into action. This may involve creating an action plan containing specific steps to be taken and a timetable for taking them. We examine these issues in more detail in the next chapter.

The Basadur Simplex Model (6)

This model again focusses on problem formulation, solution formulation and solution implementation. It comprises eight steps:

1. Problem finding
2. Fact finding
3. Problem definition
4. Idea finding
5. Idea evaluation and selection
6. Planning
7. Gaining acceptance
8. Action

The model recognises that both analytic and imaginative skills are required for success, and is accompanied by a proprietary profile tool that allows for an analysis of individuals' creative style (generator, conceptualiser, optimiser or implementer).

Team-Based Problem-Solving

As we discussed in Chapter 6, the process of creating effective teams and establishing conditions for effective teamwork encourages synergy to occur when generating new ideas. The following process is a design for establishing problem-solving groups and draws on the different models described above.Summary of the problem-solving process

Stage 1	SETTING UP THE TEAM *The initiation phase*	<ul><li>Identify the sponsoring manager to select, give authority to, and support the team.</li><li>Choose a team with the right knowledge and skills.</li><li>Gather the team, make introductions, establish roles, set ground rules. Check the team is clear about its task.</li><li>Define the problem process.</li><li>Brainstorm all possible problems.</li><li>Identify the problem worth tackling.</li><li>Clarify boundaries and agree resources.</li></ul>
Stage 2	WHERE ARE WE STARTING? *The problems phase*	<ul><li>Map the processes involved in the problem.</li><li>Work out what causes the problem.</li><li>Categorise problem barriers.</li><li>Define headers for the categories.</li><li>Prioritise the categories.</li></ul>
Stage 3	WHERE WE WANT TO GET TO, AND BY WHEN *The solutions phase*	<ul><li>What does the team want to change?</li><li>In what way?</li><li>By how much?</li><li>Over what time frame?</li><li>Brainstorm all potential solutions.</li><li>Assess the potential solutions.</li></ul>
Stage 4	HOW WE PLAN TO GET THERE *The actions phase*	<ul><li>Develop action plans.</li><li>Establish what needs to be done, when and by whom.</li><li>Share action plans.</li></ul>

Stage 5	GOING FOR IT *The implementation phase*	<ul><li>The team puts the plan into action.</li><li>Hold regular progress reviews with the sponsor.</li><li>Report out to stakeholders.</li><li>Update the plan as necessary.</li><li>Hold a final review upon completion.</li></ul>
Stage 6	WHERE NEXT?	• Identify further problems or ideas.

Stage 1—Setting Up the Team

- Identify the sponsoring manager to select, give authority to and support the team.
- Choose a team with the right knowledge and skills (see Chapter 6).
- Gather the team, make introductions, establish roles, set ground rules. Check the team is clear about its task.
- Define the problem process.
- Brainstorm all possible problems.
- Identify the problem worth tackling.
- Clarify boundaries and agree resources.

Characteristics of a suitable problem

- The problem pertains to a key area of the business or business issue.
- Tackling the problem requires a team's knowledge and experience.
- Improvement can be clearly determined and ideally measured.
- The problem can be tackled within a reasonable timescale (e.g., 4-6 weeks).
- Authority to tackle the problem, and support, are provided from the appropriate sponsor.
- The necessary resources to tackle the problem are available.

Sources of ideas

Organisations usually abound in opportunities for improvement. The challenge is to capture all these issues and opportunities such that they can be worked on. The regular trawling of ideas for improvement across the organisation, and mechanisms for sharing those ideas widely, can help to surface ideas and to prompt suggestions for improvements. Many organisations use suggestion schemes or idea management software (as discussed above), with incentives to encourage wide participation.

Idea sources within the business include:

- Bothersome problems
    - e.g., when orders get lost or are delivered too slowly
- Crises
    - e.g., no new products in development
- Business plans
- Operating reports
- Comparisons with other areas of the business (internal) or industry (external)
- Role-swaps
    - e.g., 'Back to the shop floor'. *A retailer had a competition for the best branch, and as a prize, the winning branch was rewarded with a week-long trip. For that week the board stepped in to run the branch. All the board members reflected that this was a powerful learning experience, and realised that those actions that had previously seemed to make sense in the boardroom actually meant little on the shop floor. As a result, they saw the issues and challenges of the business from a new perspective.*
- Process needs
    - e.g., over-complicated or error-prone processes that must be improved
- Customer comments/suggestions/feedback/surveys
    - e.g., changes in customer needs or demographics
- Unexpected occurrences
- Incongruities of various kinds
    - *A car company spotted a mirror taped to the driver's side sunshade in one of its cars. It highlighted the erroneous assumption (extraordinary as it seems now) that the driver would be a man, whilst a woman would be in the passenger seat and use the mirror there. On further observation of women drivers, they noticed that*

there was nowhere to put a handbag and so designed the console between the front seats to hold one.
- New technology
- Competitor pressure
- Mistakes that 'should not have happened'
    - *A company developed a range of exercise equipment and decided that it should be robustly tested. The equipment was sent to an army squad who put it through its paces, and reported back that everything worked well. The exercise equipment was then launched, but it was not long before customers started complaining that equipment was breaking. The company realised that the army squad had followed instructions when using the equipment, but customers were not so careful, leading to breakages. The company realised that they needed a better process to understand how a customer used a product in order to successfully implement further new products.*
- Staff suggestions
- Results and outcomes from previous projects

Refining the problem

If the problem…	Then…
- Is stated in terms of studying, analysing and recommending…	- Suggest a small scale and results-driven pilot to test new ideas.
- Involves resources and/or authority outside the team's control…	- Identify a smaller part of the problem that the team could tackle. - Determine whether an additional or different sponsor or team membership would be appropriate.
- Is too large and/or long-term…	- Identify sub-goals or milestones. - Work in a single geographic area. - Work with a subset of customers. - Work on one part of a production system.
- Is in an area where people are not ready to give it a go…	- Start in an area or with a problem where they are ready. - Make sure the sponsor powerfully and convincingly presents the problem or issue to the team. - Keep the sponsor heavily involved during the course of the project.
- Is so specific that it pre-empts the team's ability to decide its aim…	- Ask the sponsor to broaden its scope and/or allow the team to set specific measures and dates for accomplishment.
- Is in a department where larger changes are under way…	- Consider delaying the project or working on an issue not directly affected by the changes.
- Involves hidden agendas…	- Gently point out the ramifications of selecting this problem or issue. - Work to minimise potentially harmful impacts. - Surface the issue and facilitate the group in choosing where to focus time and attention.

Key roles

Sponsor

- Identifies the problem
- Selects the team members who, between them, have knowledge or experience of all aspects of the problem
- Briefs the team, and gives them authority (and maybe budget) to tackle the problem
- Reviews progress, provides help and support

Facilitator

- Briefs the sponsor
- Facilitates 'kick-off', the team's initiation phase
- Assists the team as needed
- Helps plan the execution of the action plan

Team

- Comprises ideally 4-8 members, including the leader/co-ordinator/facilitator
- Analyses the problem and sets goals
- Generates solutions and develops an action plan
- Implements the plan alongside the members' normal business activities

The role of team members

- Attend team meetings
- Support other team members
- Help to explore the problem and set goals
- Contribute to the generation of as wide a range of solutions as possible
- Help to create the action plan
- Take responsibility for specific items on the action plan
- Serve as liaison to others in their business department and wider in the organisation

The sponsor's brief to the team

Message	Detail
• Establish rapport with the team.	• Tell the team they are best positioned to solve problems because they are closest to those problems.
• Demonstrate the important nature of the idea or problem—to educate and to motivate the team.	• Offer compelling facts, for example: • Tell the team what a particular activity costs, and how it may affect the viability of the business unit or department. • Offer customer survey data.
• Challenge the team to improve.	• Show how the project will help team members' departments and the organisation as a whole. • Offer extreme high/low estimates of the scope for improvement.
• Authorise the team to experiment with and implement changes.	• Encourage the team to be radical. • Offer examples of potential changes the team could make that would embody real innovations or changes in procedures.
• Make boundaries clear.	• Clarify that results are to be achieved within existing staffing and resources.

Stage 2—Where Are We Starting?

- Map the processes involved in the problem.
- Work out what causes the problem.
- Categorise problem barriers.
- Define headers for the categories.
- Prioritise the categories.

Tools and techniques

Chapter 5 provides a selection of tools to use.

Analysing data

Check the data to see where, when and how it was collected. If it was a sample, establish how representative it is. Be wary of isolated events and look for trends and patterns. Compare and contrast this data with known and verified data. Be aware, however, that interpretations may differ based on how the data was collected rather than based on real differences in the data.

Balance the desirability of simple facts that are easy to understand with the danger of oversimplifying the problem. Check the logic of any conclusions, and always try to establish the root cause. (See Fishbone Diagrams in Chapter 5.)

Root cause analysis

Key questions to ask include:

- What are the facts?
    - e.g., what are the variances to the plan
    - The '5 whys' tool can be useful here
- Why have they occurred?
    - Look for the causes rather than apportioning blame
    - A fishbone analysis is a useful tool here
- When did events happen or not happen?
- How has the problem occurred?
    - What are the circumstances of the problem occurring?
    - A force field analysis can be useful here
- Where did the problem occur?
- Who did or did not do what?

Stage 3—Where We Want to Get to and by When

- What does the team want to change?
- In what way?
- By how much?
- Over what time frame?

- Brainstorm all potential solutions.
- Assess the potential solutions.

Agreeing the statement of aim/goal

- Agreeing the goal of the team is the most important part of successful problem-solving.
- A clear, shared, measurable goal turns a group into a team.
- The team needs to agree as specifically as possible:
    o What is the change we want to make?
    o Over what period do we want to make the change?
    o In what area do we want to make the change?
    o By how much do we want to make the change?
- Ask the team for a list of possible changes and narrow them down, using questions such as:
    o Is the change measurable?
    o Is there one change that would incorporate the others?
- Aim to build consensus in the process of choosing the change to focus on.
- Check to see if your goal is SMART
    o S—Specific and stretching
    o M—Measurable
    o A—Attractive and attainable
    o R—Realistic and relevant to the team
    o T—Time-framed

Stage 4—How We Plan to Get There

- Develop action plans
- Establish what needs to be done, when and by whom
- Share action plans

Drawing up a plan of action

Ask the team to spend 5 minutes individually writing down all the things each member thinks must be done to achieve the goal the team has set. To help them, they can refer to the goal, the 'typical' elements of the plan, the problem analysis and the process map if they have one.[19]

Invite each team member in turn to give you their ideas. All ideas should be accepted and recorded (e.g., written on a flip chart) without judgement or evaluation. If people think of more actions during this process, they should add them to their list.

Make sure the list includes key actions such as, 'nominate team co-ordinator for implementation', 'write up and circulate action plan', 'present plan to sponsor'. Evaluate the list of actions against the statement of aim; if an action is not likely to contribute directly to achieving the goal, ask if the team still wants to do it. Highlight all actions that the team wants to pursue.

Next decide on the start and end dates for each action, ensuring that all actions are carried out in time to meet the action plan deadline. Some actions will depend on others—the team may not be able to start one action before another has been completed. Using a Gantt chart (see appendix) can be very helpful in this process.

Finally, invite the team to decide which team member will be responsible for delivering each action. This does not mean that they have to do the action themselves, but it is up to them to see that it is done. Use a chart, like the one below, to track roles and responsibilities.

Action	Start Date	Finish Date	By Whom

Stage 5—Going for It

- Put the plan into action
- Hold regular progress reviews with sponsor
- Report out to stakeholders

---

[19] A typical plan might consist of: a list of activities to be done, responsibilities and timelines for each activity, resources required including budget authority, communication with key stakeholders, risk analysis and implementation strategy.

- Update the plan as necessary
- Hold a final review upon completion

How teams feel

The graph below plots the changing level of confidence experienced during a project over time and illustrates what a roller coaster it can often be.

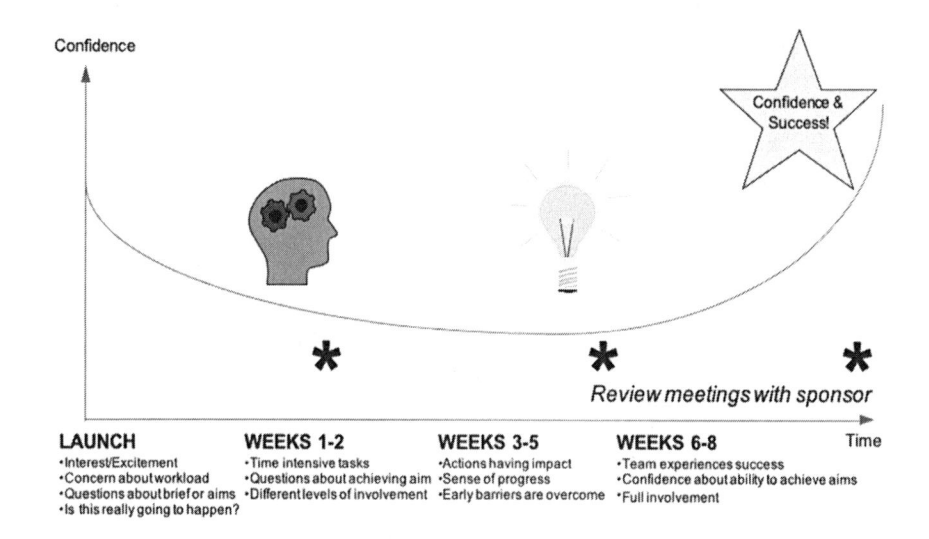

	LAUNCH	WEEKS 1-2	WEEKS 3-5	WEEKS 6-8
	•Interest/Excitement	•Time intensive tasks	•Actions having impact	•Team experiences success
	•Concern about workload	•Questions about achieving aim	•Sense of progress	•Confidence about ability to achieve aims
	•Questions about brief or aims	•Different levels of involvement	•Early barriers are overcome	•Full involvement
	•Is this really going to happen?			

Review meetings with sponsor

Timetable for reviews

Typical review pattern for an 8-week project:

Week 0 Launch
Week 2 'Team review' meeting with sponsor
Week 5 'Team review' meeting with sponsor
Week 8 Final review and celebration

Regular reviews with the sponsor demonstrate a high level of sponsor engagement and involvement, which is highly motivating for the team. Regular reviews also give the team milestones to aim towards, and enable the team to formally update the plan in the light of new information and events that surface in the review process.

The following structure can be helpful for presenting at a review meeting (7):

- Outline the problem that has been tackled
- Show the initial analysis
- Explain the reasoning behind the data that was collected
- Present the facts that were found
- Go through all the possible solutions
- Show why the chosen option is preferred and why others were rejected
- Present the benefits that will accrue from this solution
- Detail the implementation plan and any risks
- Open the presentation to questions and discussion

In preparation for a presentation, two techniques can be useful to counter the potential problem of group think. One approach is to play 'devil's advocate,' where a person is designated to deliberately challenge the idea and look for all the possible flaws. Another approach is to establish 'a red team' to identify alternative options (discussed in Chapter 8), which can sharpen the thinking.

Stage 6—Where Next?

- Identify further problems or ideas.

Chapter Summary and Conclusions

This chapter has examined several models and processes for engaging groups in creative problem-solving work. Facilitation of these events requires finding some balance between a disciplined process and sufficient flexibility to incorporate unplanned discussions.

Approaches covered include:

- Points to consider in setting up a group process
- A generic problem-solving process
- The Osborn Parnes Creative Problem-Solving Model
- The Basadur Simplex model
- The 6 steps in team-based problem-solving

Chapter 8
From Creative Spark to Acceptance

Introduction

Winston Churchill, prime minister of war-torn Britain, once remarked, "However beautiful the strategy, you should occasionally look at the results." (1)

In this chapter, we will discuss the journey that leads from ideas to results. We will identify some of the elements, techniques and activities that facilitate the process of turning an idea into a practical change that brings benefit.

We approach this topic from the perspective of the individual seeking to introduce their idea and creative work into the world, and from the perspective of the organisation that is trying to make it easier for ideas to be implemented. In doing so, we pull together several of the findings identified so far through this book into a coherent strategy for action. The chapter is structured in six sections:

1. Process
2. Persuasiveness
3. Confidence
4. Commitment
5. Practicality
6. Working with resistance

In 1870, the wife of a miner complained to Jacob Davis, a tentmaker who worked with cotton duck, twill and denim, that her husband's work trousers were constantly tearing. Inspired by the sight of the copper rivets on his bench that were usually used for securing tents and wagon covers, Davis made up some overalls that were strengthened with the rivets at the pockets and seams, and the Jean was born. These trousers were made with denim supplied by Levi Strauss, and when Davis applied for a patent in 1873, he and Strauss went into business together manufacturing jeans (2).

This story illustrates several characteristics of the successful implementation of an idea to generate positive benefits: a problem that required solving, the quest for a solution that provoked enquiry, materials from other applications that could be used for the new application, and existing organisational infrastructure (the respective businesses of Davis and Strauss) to facilitate production and sale.

As we have discussed throughout this book, there is, of course, a problem with the 'creationist' view of innovation; i.e., in a genesis-like moment a new idea springs to mind, leading to new products straight from the drawing board.

The idea ('inspiration') is simply the start of the journey, and it's the significant amount of perseverance and 'perspiration' that will actually determine whether the idea will translate to practical success and benefit or not. It is this process of harnessing perseverance and perspiration to bring the idea to life that we will discuss in this chapter.

Reflecting on the story of Chester Carlson and the invention of xerography (the process of photocopying), Joe Wilson commented, "If we had fully foreseen the magnitude of the job (i.e., the millions of dollars needed for research, the marketing complexities and the manufacturing difficulties), we probably would not have had the fortitude to go ahead. But we did" (3). This chapter discusses a number of approaches to facilitating and dealing with this 'magnitude of the job' in any creative endeavour.

1. Process

What's the problem?

I didn't create these businesses to make money, I created them to solve problems. – Muhammad Yunus, Bangladeshi social entrepreneur (4)

The Jean was born from the desire to solve a problem, and it's this quest to solve a specific problem that commonly provides a source of motivation for creativity and innovation. It is a powerful driver of research, experimentation and enquiry.

Sometimes, creative thinking and research are not aiming to solve specific problems but instead lean towards self-expression and an expansion of knowledge based on curiosity.

However, for creative ideas to lead to practical benefits, a focus on finding a solution to a problem is often required. The organisations that have most success in bringing creative ideas to life have given considerable thought to the process by which they do so. How do they facilitate the creation of ideas, and the journey from idea to practical outcome?

Professor Robert Langer, a chemical engineer whose Langer Lab conducts research at MIT's Koch Institute, has been highly successful in generating new ideas and bringing them to market as commercial products. Nicknamed 'the Edison of medicine', he holds over 1,100 current and pending patents, and findings from his lab have led to the creation of some 40 companies with a collective estimated market value in the billions of dollars.

He describes five elements to his approach at Langer Lab: "A focus on high impact ideas, a process for crossing the proverbial 'valley of death' between research and commercial development, methods for facilitating multidisciplinary collaboration, ways to make the constant turnover of researchers and the limited duration of project funding a plus, and a leadership style that balances freedom and support" (5).

His highly successful approach in developing new medical products starts with a focus on projects that will have a major impact on society. ("Do something that's big. Do something that really can change the world rather than something incremental.") He reasons that market demand and financial benefit will follow more easily from major impact innovation than from small incremental improvement.

This focus on radical ideas defies rational budgeting and cost benefit analysis, and demands a high-level commitment through the many ups and downs and obstacles to the research, realities that companies often shy away from. It requires a clear understanding of what you say no to, and an awareness of your mental filters. It also requires an acknowledgement of the need to focus on only a few things and to do those few things well.

A summary of the Langer Lab approach is to:

- Pursue use-inspired research (i.e., research that aims to address and resolve practical problems)
- Nurture deep scientific and engineering expertise in a handful of areas
- Manage intellectual property aggressively
- Treat the central research organisation as a separate entity, liberated from the incremental demands of established business units
- Staff laboratories with great—not merely good—scientists and engineers, with an emphasis on making a difference rather than job stability
- Establish consistency over time in the funding of, organisational approach to, and independence of, advanced research units
- Ensure robust leadership

Reinforce the Innovative Culture

As we have discussed elsewhere in the book, challenging existing assumptions and ways of doing things can stimulate the rethinking of current patterns of behaviour. This can occur when an independent voice joins the process. Companies often have induction programmes based on the view that 'our ways of doing things are correct and you can forget what you learned in a previous company'.

One financial institution flipped this mentality by giving new recruits a list of people they should talk to in order to find out about the company. They asked them to note down anything which seemed strange (or not very sensible) and return in a fortnight to present the findings. They discovered a wealth of small improvements in all aspects of the business that had been overlooked due to the acculturation that quickly occurs to employees in the business.

It is the task of leadership to craft a challenge and a vision that energises and inspires and gives freedom to explore. This is more likely to be successful when this challenge and vision resonate with both the organisation's current problems and their prior experiences. It is easier to pursue ideas that fit with your existing mental framework, and indeed with the mission of your organisation, than to redefine your craft. However, this can lead to myopic thinking.

Tools manufacturer Black and Decker initially turned down the Workmate, a portable workbench and one of its most iconic and successful products, because it didn't have a motor, and Black and Decker had defined itself solely as a manufacturer of portable powered products. By redefining its business as a manufacturer of products that help DIY and tradesmen, regardless of whether they were powered or not, and issuing a challenge to seek out new products, the company unleashed a wave of new thinking and a wider search for ideas.

Netflix similarly excelled when they redefined their business from mail-order DVD rental company to TV content creator and online distributor. Parker, the pen manufacturer, successfully redefined their business as a gift company rather than a writing company and placed greater emphasis on the styling and presentation of their pens as a result.

This process can require a balancing act that is challenging to maintain: an organisation must break out of familiar (and failed) parameters of thinking and behaving in order to redefine its business, but must make sure not to spread itself too thinly, as this can lead to a squeezing of resources.

One of the challenges of conducting research, particularly in its early stages, is managing the risk—of wasted time, money and resources. One way of managing this risk, particularly when investing in early-stage research, is to look for areas of research that might have a number of applications, as well as adopting an openness to pursuing unintended uses that might arise.

Research conducted by Momenta, a company using Robert Langer's research into the structure of sugar molecules to focus on cancer treatment, led to an application for treating deep vein thrombosis. Another renowned unintentional discovery was that of Viagra, which was initially developed for hypertension, but subsequently found a lucrative market in the treatment of erectile dysfunction.

In Chapter 6, we identified a number of features of a culture that supports innovation. Managers who have the responsibility for encouraging innovation can do several things to facilitate the development of new ideas and the implementation of new and innovative products.

With an aim to reinforce an innovative culture, Dave Allan, Matt Kingdon, Kris Murrin and Daz Rudkin of the innovation company? What If? suggest the following practices in their book *Sticky Wisdom* (6). Many of the suggestions align with the Idea Generation Techniques described in Chapter 5.

- Get into different ways of thinking by:
  - Finding alternative ways of describing something, using the language of a different discipline.
  - Using a framework: e.g., replace, combine, minimise, maximise and/or adapt existing products.
  - Finding an alternative in another world.
  - Deliberately challenging the rules and assumptions.
  - Looking for random connections.
- 'Green housing'
  - Take a SUN approach: Suspend judgement, Understand ideas and differences and Nurture embryonic ideas.
  - Avoid a RAIN approach: React, Avoid and INsist.
- Realness
  - Try to make ideas as real as possible
  - Aim for prototypes, mock-ups and simulations (which can reduce the risks if they fail and avoid jeopardising the entire project).
- Momentum
  - Get something going—this avoids getting stuck in the inevitable inertia that organisations suffer from. It also encourages experimentation which can be valuable because not everything can be anticipated in theory.
- Signalling
  - Indicate to others what you want—support, advice, criticism.
- Courage
  - Be true to yourself, and the courage to stand up for what you believe in will follow.

Creative output is often characterised by a deep knowledge and expertise in a specific field, usually borne out of a passionate interest and sheer hard work on the part of team members to master their craft (7). Organisations and individuals that continue to invest in their talent (and give freedom to use it), both through formal learning and through informal opportunities to learn from others, will be more likely to succeed.

Finally, we should recognise that creativity does not just come from a logical process. "Creativity is a process that reflects our fundamentally chaotic and multifaceted nature. It is both deliberate and uncontrollable, mindful and mindless, work and play." (8) Any processes, structures and activities that are introduced to facilitate creative outcomes must also be prepared to bend and give way to these chaotic forces. The balance between process and chaos can be tense.

2. Persuasiveness

For any idea to progress to a practical outcome, the originators of the idea must enlist support—from colleagues, managers, investors, customers, retailers and so on. Their capacity to persuade these stakeholders of the value of their idea is paramount to its flourishing. And this is no easy task; resistance will inevitably be challenging. Creativity and innovation, the enemy of stability and conformity, almost by definition, imply change and naturally invite resistance from those who are currently stakeholders in the status quo.

In his 1514 book *The Prince*, the Renaissance political philosopher Machiavelli famously wrote (9), "There is no more delicate matter to take in hand, nor more dangerous to conduct, nor more doubtful in its success, than to be a leader in the introduction of changes. For he who innovates will have for enemies all those who are well off under the old order of things, and only lukewarm supporters in those who might be better off under the new. This lukewarm temper arises partly through the fear of adversaries who have the law on their side, and partly from the incredulity of mankind, who will never accept anything new until they have seen it proved by the event."

Persuasive influence is key, and much has been written about the art of persuasion, from Dale Carnegie's seminal book *How to Win Friends and Influence People* (10) to Roger Fisher and William Ury's book on negotiation, *Getting to Yes* (11).

A common misconception about influencing is that it is best achieved by using logic to speak to an audience. The Greek philosopher Aristotle suggested that we communicate in three ways: with logic (Logos), with emotion (Pathos) and with our credibility in the topic we are championing (Ethos). We can use all three approaches to increase our capacity to influence and persuade our audience.

Furthermore, it is important to recognise that persuading someone is not a one-way street. St Francis of Assisi is credited with the expression: "If you wish to be understood, seek first to understand." This sentiment was later popularised by Stephen Covey as one of his 7 habits of highly effective people (12). By understanding what motivates your audience, you will gain valuable insight into what might persuade and influence them.

What these quotes also illustrate is that influence is about establishing connection and enabling a relationship of equality to emerge. Accordingly, research has shown that people who score highly on agreeableness as a personality trait are more adept at getting people to listen to their creative ideas.

More recently, the field of positive psychology has been influenced by the 'Nudge' movement, now widely adopted by certain governments in their attempts to shift public attitudes and behaviour. Richard Thaler and Cass Sunstein popularised the approach in their book on the topic (13), and they describe a series of strategies of 'choice architecture' which aim to encourage wiser behaviour.

The underlying assumption is that by giving people choice, but using positive reinforcement and suggestion, people will make better decisions than if they were forced to take a specific action. NUDGES is a helpful mnemonic for the process of nudge strategy: iNcentives, Understand mappings, Defaults, Give feedback, Expect error and Structure complex choices. Let's explore each of these:

Incentives

Incentives to adopt new ideas must be stronger if financial gain for the people you are aiming to influence is not an option. This is based on the economic view of humans (which is not always true) that money is the most powerful incentive.

A good starting point in the investigation of incentive is to employ the questions: who uses, who chooses, who pays, who profits? The answers to these questions are simple in the case of a music fan purchasing a new music recording from a retailer. The purchaser chooses a recording, assesses the price, pays for it with their money and enjoys listening to it, while the costs and likely profit are known to the seller. It becomes more complicated in the case of an artist manager who is pitching a new band to a recording company.

The person within the record company who acts as the initial gatekeeper (e.g., a junior employee or intern who is listening to all the demos) may have no stake in the band's success and is without influence on those who make the final decision or who bear the financial consequences of any choice.

Social media has changed the balance of power in many industries, particularly the arts, and provided the opportunity to reach a wider audience who might have greater influence on the final decision maker. The 'early adopters' of an idea may well be spotting a need ahead of the wider market and should be watched carefully.

Throughout history, financial rewards have been used to incentivise creative experimentation. As a recent example, the Ansari X Prize was a space competition in which the X Prize Foundation offered a US$10 million prize for the first non-government organisation to launch a reusable manned spacecraft into space twice within two weeks. It was modelled after early 20th-century aviation prizes (including the one that incentivised the very first motor-powered flight achieved by the Wright brothers) and aimed to spur development of low-cost spaceflight.

Understand Mappings

The concept of mapping is based on the idea that by presenting information in a format that is easier to understand, people can make better choices. Too often information is presented in ways that make meaningful comparison extremely challenging, if not impossible—like the proverbial comparison between apples and oranges. Furthermore, the way that information is framed can have a large impact on its influence. Consider the following example.

When patients were considering treatment options for an aggressive cancer, they were presented with the following statistics:

- Option A: Of 100 people having surgery, 10 will die during surgery, 32 will die within one year, and 66 will die within 5 years.
- Option B: of 100 people having radiation therapy, none will die during treatment, 23 will die within one year and 78 will die within 5 years.

Which option would you prefer?

If the options are now framed in a different way, what would you now choose?

- Option A: Of 100 people having surgery, 90 will survive surgery, 68 will survive past one year, and 34 will survive past 5 years.
- Option B: Of 100 people having radiation therapy, all will survive treatment, 77 will survive one year, and 22 will survive past 5 years.

When this question is presented to various audiences using the first framing, there is a 50:50 split between the two options. When the second framing is used, over 80% of people choose Option A, the surgery. Whilst still a difficult choice, mapping the two options and framing them in such a way as to focus on long term survival rates makes it easier to make the choice.

When seeking to influence opinion-formers about a new product, a focus on the features that resonate most with consumers is key. Car advertisements highlight the lifestyle they facilitate rather than discussing the improved titanium catalytic converter, although when targeting a niche market such as environmentally conscious consumers, the latter may be just the feature they want, especially when the discussion is accompanied by independent comparison data showing lower pollution levels.

Apple has changed the economic view of innovation adoption by creating a powerful community of dedicated users. Apple's brand values and designs elicit in potential customers a desire to be part of the Apple community, not just to buy a product.

There is an inherent bias in most people towards established practical solutions over novel and untried ones. Practical application to the audience must be the focus.

Defaults

In our busy lives, we tend to take the path of least resistance unless we are passionate about something. The principle of Occam's razor, whereby the burden of proof of adopting a new theory or idea favours the simplest solution or the one that makes least assumptions, is a powerful factor.

Banks, insurance companies, utility companies all report that the switching rate between them is low, because it is a hassle to switch to another provider. Indeed, the profits of many of these companies are predicated on this unwillingness to switch, even as the companies raise their prices year on year. The process of pitching a new idea is met with a similar inertia as people find it easier to simply stay with the status quo.

In the diagram below, the performance of two products is tracked over time. The first curve illustrates the performance of the original product, the second curve illustrates that of the new/updated product. The challenge for organisations is to launch the new product at the right time.

If the organisation waits until the point when the original product's performance starts to decline, it is already too late. There is no time for the launch, experimentation and testing of new products and services, let alone time for them to gather momentum and increase performance, in order to maintain the overall performance levels of the original product.

There are also no resources available—instead, the name of the game is usually survival for the original product, and all resources and efforts are focussed on trying to boost the performance of the original product, which is starting to decline.

Actually, the ideal time to launch the new product is at the top of the performance curve of the original product, or even earlier (as shown on the graph—and critics will rightly point out that it is a bit theoretical). This provides the necessary time to launch, experiment, test and build momentum, so that when the performance of the original product starts to decline, the new product's performance is beginning to surge.

However, even launching the new product at the right time—where the second curve on the graph starts—brings its challenges. There is still a fight for resources but for a different reason: 'if it isn't broke don't fix it'. It can be tough for organisations to funnel resources into launching and developing the new product when the original product is doing so well (and continuing to increase in performance), since the temptation is to funnel everything towards the original product in order to maximise its performance.

The moral of this phenomenon is that the case for change must be stronger than staying with the status quo (see the burning platform argument later), and where the practical relevance of the curve below can be demonstrated, the case is enhanced (14).

The challenges of change

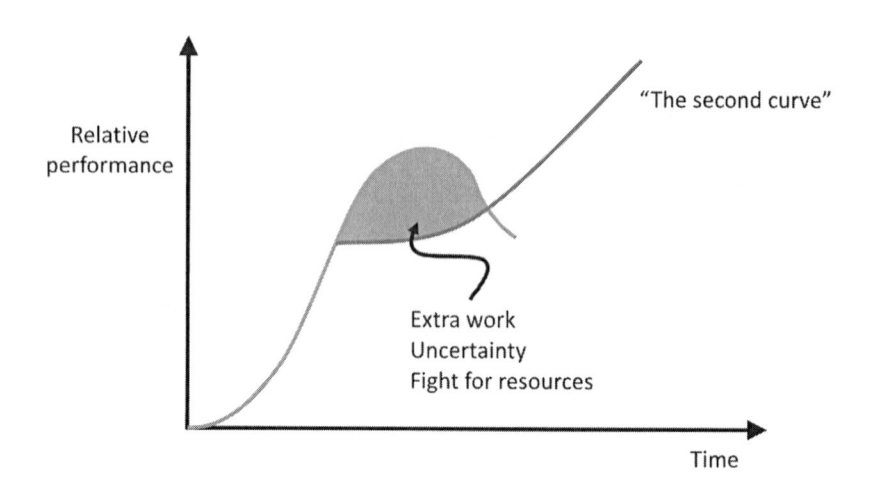

Relative performance (y-axis) vs *Time* (x-axis)

"The second curve"

Extra work
Uncertainty
Fight for resources

Copyrighted by Brain Technologies Corporation of Gainesville, Florida, and reproduced from the book, *Strategy of the Dolphin*(tm): *Scoring a Win in a Chaotic World,* by Dudley Lynch and Paul L. Kordis, William Morrow and Company, 1988, p. 79, and other works.

Give Feedback

"Feedback," said management guru Ken Blanchard, "is the breakfast of champions."

It is difficult to improve performance if you don't have feedback; feedback provides the critical information required to make the necessary changes to improve—change what didn't work so well, do more of what did work well, and so on. Feedback can be given and received in a number of ways—one-on-one conversations with a manager or team leader, analysis of data such as sales figures, a post-mortem after an experiment, event or product launch.

However, the process of giving and receiving feedback can be very challenging. One of the major issues is that feedback is often mistaken for personal criticism—a critique of the person, of who they *are*, not of what they *did*—and this can cause angst and be demoralising.

Remember the key components of effective feedback: feedback is intended to be beneficial for the recipient, it should be based on factual and preferably first-hand observations of behaviour, it should highlight the impact of such behaviour, and finally should focus on taking effective action to move forward.

As well as discussing 'feedback' it can be helpful to consider the concept of 'feed forward'. Aim to balance the process of deconstructing what happened in the past with a positive perspective on how the learnings from those experiences can be used constructively in the future. If feedback is focussed solely on what happened in the past, without a view to the future, it can be destructive and extremely frustrating and demoralising, since it just picks over history and history can't be changed.

This past-focus can elicit fear and insecurity about competence and feelings of rejection. However, the point of studying the past is to learn for the future, so there must be focus on how these learnings can be applied moving forward. The concept of 'feed forward' is about ascertaining what we can learn from the past to be better in the future.

Several factors help us to deal with the criticism that can be implied in feedback. Grit, resilience, and mental toughness—factors that we discussed in Chapter 4—can be extremely beneficial here, not only in taking feedback constructively, but also in persevering with the creative task and implementing the learnings moving forward, rather than giving up at the first sign of failure.

Reframing any negative feelings that arise, acting 'as if' to change the self-talk, and recalling past successes and positive feedback can all help to restore confidence, learn from feedback, and persevere towards creative goals.

In particular, the support of others who believe in us can be vital. We might not have had Van Gogh's paintings but for his brother Theo who supported him financially and emotionally. This role of supportive friend can have its challenges, however; it may require speaking the truth and giving honest feedback, but also providing a boost to self-esteem and encouraging the belief that the work will (at some point) yield positive results.

Expect Errors

The capacity to normalise failure as a common part of the creative process must be learned early in the creative journey (remember Henry Ford and Thomas Edison).

A common issue that can arise is a failure to fully complete the job in hand. A complaint that is often heard is: "Every DIY job in this house is only 90% complete. Why don't they finish the job?" The reason is often boredom: the enjoyment is in the process of dreaming up the creative idea—music, painting, a new gadget, a life hack, a lifesaving drug—and the hard work that follows to actually successfully implement the idea is often a drudge. Post-completion errors can also arise, particularly if the motivation to be detail-oriented when undertaking the hard work starts to wane.

Structure Complex Choices

If we present people with a large array of choices, they tend to adopt simplifying strategies to narrow down the number of choices using simple criteria to eliminate options, a process called elimination by aspects (15). The danger with so many choices is that they can actually become so bewildering as to prevent any choice from being made at all.

The anxiety and paralysis that can occur for people in this situation is part of the paradox of choice, and by eliminating or simplifying choices, we can make the choice process easier and less stressful. Paint manufacturers have developed a huge range of colours (often with bizarre names that are a struggle to connect to any colour).

However, they have then simplified the process of selection through the use of colour cards, which provide just a small number of choices for an overall colour-type, as well as a visual reference tool for those options.

3. Confidence

"As is our confidence, so is our competence," said William Hazlett, an 18th century philosopher. Creative people often display characteristics of self-doubt, especially at the time when their new creation goes public. Painters sometimes destroy their work, engineers abandon their projects, musicians shelve their projects, all as a result of doubt about its value.

However, the confidence (or lack thereof) that we project in our creative ideas and endeavours can have a large impact on our capacity to persevere with them, as well as for others around us to accept them. The influence on outcomes that our confidence levels can have is encapsulated in a well-known poem by Walter D. Wintle (16).

If you think you are beaten, you are;
If you think you dare not, you don't.
If you'd like to win, but you think you can't,
It is almost a cinch you won't.

If you think you'll lose, you've lost;
For out in this world we find
Success begins with a person's will
It's all in the state of mind.

If you think you're outclassed, you are;
You've got to think high to rise.
You've got to be sure of yourself before
You can ever win the prize.

Life's battles don't always go
To the stronger or faster man;
But sooner or later the person who wins
Is the one who thinks he can!

However, an abundance of confidence can have its drawbacks too. One challenge is the planning fallacy which prompts us to take an overly optimistic view of the time and resources required to deliver a project.

Collaboration and partnership, whereby participants bring different skills and abilities to the table, can be a useful way to address these confidence issues. On the one hand, collaborators can provide a boost of support to each other's confidence; on the other hand, they can provide a voice of reason and realism in order to keep the planning fallacy in check.

4. Commitment

After a performance, pianist Richard Clayderman was visited backstage. The visitor said, "I would give my life to play the piano like that."

Clayderman replied, "I did."

Such commitment ensures that the necessary perspiration for creative outcomes takes place, and also elicits a level of admiration and respect that often means the creative work is taken seriously. It is an essential quality of successful artists and scientists alike. Michelangelo reportedly worked day and night when working on his masterpiece sculpture *David*, refusing to sleep or eat unless he was forced to.

Reflecting on the key to his own commitment through the words of one of his characters, Charles Dickens wrote, "I never could have done what I have done without the habits of punctuality, order and diligence, without the determination to concentrate myself on one subject at a time" (17).

This production phase is critical to success. However, it is easily hamstrung by distractions, a lack of skill or resources, a lack of self-discipline, and crippling self-doubt, while intrinsic motivation and a passion for the potential outcome will aid the process. In Chapter 4, we discussed a variety of approaches to these challenges.

Commitment is also crucial from those who have control of the resources and authority required for successful implementation of a creative project. Michael Ilczynski, founder of the job-seeking website *Seek*, hosts an annual 'hackathon' where groups come together to work on current problems. "There are no rules, no designated themes and no vetting or censoring by management," he says. The exercise is competitive and, crucially, there is an absolute commitment that the winning team's proposal will be moved into production. A third of the solutions that are generated are internally focussed and designed to lead to a happier and more efficient workplace (18).

Commitment is also affected by several characteristics:

Personality and Production

The name of 'madman' with which it is attempted to gag all innovators should be looked upon as title of honour (19).

Creative thinking, and the pursuit of creative success, can take people on a rollercoaster of emotional highs and lows of mood. As a result, the reality of an idea's production phase might not always adhere to the expected plans and timetables.

Faced with the contrariness that can arise when people express different viewpoints of the world or are reluctant to accept the norms and constraints of others in authority within organisations and society at large (all hallmarks of creative idea generation), delays are a common outcome. It doesn't help that the clichéd Bohemian lifestyle of the artist without regard for deadlines is often true.

Ludwig (20) identified several other characteristics which may hinder the creative process. The desire for solitude, and a desire for dominance (think of Steve Jobs at Apple) can easily result in friction when people are required to work in teams and on committees.

Thinking Biases

Kahneman and Tversky (21) identified a series of 'heuristics and biases' which illuminate human judgement. For example, we tend to anchor our thinking by our most recent experience—what we have just heard, read or seen. If a selection of people were asked to guess the size of the population of Turkey, their answers would be more scattered in range than if they had been asked whether the population was more or less than 80 million and by how much more or less. In the latter case, people's thinking has been anchored by the mention of '80 million' and they are more likely to choose a number that doesn't stray far from this anchor. (Incidentally, the population of Turkey is roughly 80 million people).

Accordingly, a novel idea is more likely to be taken seriously if the audience has had the experience of crazy ideas being implemented before. However, if the audience is more familiar with examples of ideas that haven't worked, they will have a greater negativity bias, that is, they are more likely to look pessimistically on the chances of success. (Remember however, the attitudes of Ford and Edison to failure, encapsulated in the quote: *I haven't failed, I just know what doesn't work*).

We are also prone to make judgements based on stereotype, what we think someone or something *should* be (the representativeness heuristic). This is often encapsulated in well-worn phrases, such as the 'absent minded professor' or the 'drug-fuelled musician' or the 'tortured painter or poet'. If a creative idea does not fit with the stereotypes we know, we may be biased against that creative idea.

When forecasting and making estimates we tend to extrapolate from the past. This is a common failing in investments. We buy at the top of the market (because the past tells us that the price will keep on going up) and we sell at the bottom of the market (because we believe it is going to get worse still). Those who make money do the reverse.

We are greatly affected by 'group think'. In a famous experiment, a group of people were ushered into a room where someone was already sitting. The lights were turned out and another light shone at the end of the room. The group was asked which way the light at the end of the room was moving. Soon someone said 'left, up, right, down...', repeating the pattern of movements, and soon most of the group were joining in.

In fact, the light at the end of the room had never moved. The person already sitting in the room, who had started calling out the movements in the dark, was a plant. Since they had spoken with confidence and assuredness, and because the group was in the dark with no other clues or data to go on, everyone else had followed their lead. People are often afraid to speak out against confidently stated opinions in case they are ridiculed, or against the prevailing communal belief for fear of being ostracised or fired.

'Red teaming' and playing 'the devil's advocate' are both mechanisms for rigorously challenging widely held or institutionalised beliefs. By employing these mechanisms, organisations can robustly assess any group think or other faulty thinking without the issue becoming a personal one.

Given that we are also biased against unknown risk, and have a tendency to take a critical approach to anything new or unfamiliar, it is clear why a new idea might be rejected by a committee if one person speaks authoritatively against it (whether they know anything about the subject or not!).

Committees can have a political dimension with competing factional interests, leading to viewpoints being expressed that are self-serving rather than objective. People can also be motivated to speak critically about ideas and projects by wanting to project themselves to others as a certain persona (wise, knowledgeable, etc.)

All these thinking biases can impact our commitment to persevering with ideas; no less importantly, they can also influence the extent to which those around us will commit to supporting our efforts and our ideas. This commitment from others may make all the different between success and failure, so we must try to identify the thinking biases that might be in play, and consider how we might navigate them accordingly.

So, let us turn to what can be done practically to overcome some of these hindrances.

5. Practicality

Of all the ideas and opportunities that emerge, how do we establish a process to decide what to invest in and pursue further? Dragon's Den, a popular TV show, has a particular approach: those with ideas are invited to pitch to seasoned investors, and after a thorough grilling, the investors decide whether to invest in the idea or not. This process requires several ingredients to run successfully: framed questions that get to the heart of the practical application of the idea, sufficient supporting data for each idea, transparency in the underlying technology of any idea, market research to demonstrate its scalability, a set of scorecard metrics to evaluate the idea, and judges who will back the idea with (ideally) their money or reputation.

A useful approach is to use collaborating groups to enhance each other's ideas. For example, two groups each take ideas from the other and seek to improve upon them and turn them into practical solutions. The process can be competitive with multiple groups competing to develop the most practical solutions from the initial ideas.

Change Management

As we discussed earlier, creativity and innovation are the enemy of stability and conformity, imply change, and may well encounter resistance from current stakeholders in the status quo. Bringing about change through the implementation of new ideas is a difficult process to execute.

John Kotter's book *Leading Change* (22) explores a useful set of considerations for any process of change. We examine its application to the adoption of innovation below.

The Burning Platform

Establish a sense of urgency for the change to take place. This is sometimes described as 'the burning platform' that you have to find a way to get off. Tales abound of inventors and artists of every kind who spent long periods of time persuading and cajoling people to give them a chance. Those who are the guardians of the status quo have little incentive to take a risk unless they are sufficiently captivated by the proposition or there is a compelling reason to act now. Never underestimate the power of inertia.

Numerous performers have got their lucky break only because of a last-minute cancellation. It is often a crisis that prompts inventors to solve a specific problem ("You never let a serious crisis go to waste"—Rahm Emanuel), and the solution often has value beyond the crisis. New approaches to working, socialising and home-schooling have been pioneered in response to the Covid 19 pandemic, some of which will endure after the pandemic subsides.

As the phrase goes, 'necessity is the mother of invention', and it is no surprise that war is a major driver of innovation of all kinds, not just technical but social as well. The development of Colossus, the first computer, to aid in code-breaking at Bletchley Park during the Second World War is a case in point.

The recent Covid 19 pandemic has thrown up many examples of how businesses have had to reinvent themselves to survive an economic crisis. Companies are collaborating in drug development where previously they jealously guarded their research. Sysco, an American food distribution company focussing on restaurants, built a new supply chain for grocery stores within a week.

Stagekings, an Australian company that built stage sets for live performances, festivals and events, used its expertise in working with wood to create a range of easy to assemble/disassemble desks and other work-from-home furniture. Speed, a lack of bureaucracy and a focus on experimentation rather than analysis are the hallmarks of this period of innovation.

Companies often use deadlines and demanding goals to create a sense of urgency, whilst the threat of being left behind by the competition represents the 'burning platform' that provides a spur to finding that successful new product.

Guiding Coalition

There is an anecdotal rule of thumb that about a third of the people affected by an innovation need to be in support of that innovation for it to take hold.

Another third might be adversely affected (perhaps because they have a stake in the existing situation), while the final third are the 'floating voters' who might be swayed either way to support or reject the innovation. But with one third on side, there is a greater chance of persuading the floating voters, meaning that successful adoption of the innovation is more likely.

A common strategy is to seek key opinion formers and use them to persuade the floating voters. This creates the guiding coalition that will spearhead the acceptance of innovation and change. In organisations information can spread through both formal communication networks and more informal rumour routes.

At nodes in these networks you may find 'telephone exchanges'—people who seem to be connected to everyone, know what is going on, and are well placed to influence opinion. These 'telephone exchanges' are key to persuading others, and in particular, to getting the 'floating voters' onside.

A water utility company was implementing a large change programme but faced resistance to the change. They sent a research team to different countries to study the latest ideas in water treatment; this team included people in a range of roles throughout the organisation, from shop stewards to those who had influence at all levels of the organisation. The team returned from the trip brimming with ideas, and because the team members came from all levels and areas of the organisation, they were able to encourage support across the organisation for the change programme.

Vision and Strategy

Develop a vision and a strategy. People buy in to a vision if it is accompanied by a strategy for implementation. The challenge can arise that the evolution and implementation of a change programme means that the existing strategy is no longer applicable. As a result, it is imperative to establish a new vision, and communicate the new vision throughout the organisation by every means possible.

Whilst the burning platform is a powerful catalyst for action (a situation to escape from), a more positive future vision (a goal to aim towards) will help to sustain effort in the future.

Empower People

Giving people a stake in the process of change ('skin in the game') builds motivation and commitment. It is rare that creative people, whether scientists or performers, are there solely for the money. Their motivations have much to do with their craft and their curiosity and drive.

Indeed, ensuring the originators of ideas are involved in bringing the product to market helps drive the process, ensures commitment and encourages learning from the harsh commercial realities of the process. In the UK, NESTA (the National Endowment for Science Technology and the Arts) places a major emphasis on the involvement of originators in implementing new ideas.

It is important, however, that this 'skin in the game' applies to the outputs of the creative process as well as the inputs. Royalty agreements from publications and patents are crucial to maintaining the flow of ideas in the future. MIT (the Massachusetts Institute of Technology) ensures that those responsible for successful invention receive one third of the corresponding royalty income.

Change that is done *to* people invariably arouses resistance, whereas change that is done *with* people elicits greater buy-in, as well as providing the opportunity to shape new innovation with more numerous and diverse inputs. The New Economics Foundation (NEF) has pioneered a process of co-production with numerous stakeholders in their aims to accelerate the adoption of new ideas in the public sector.

Generate Short-Term Wins

The conundrum of any innovation is that it is difficult to prove to sceptics the innovation's impact and benefit without concrete evidence, yet it's only possible to undertake the experimentation that provides this evidence with the support of the sceptics. 'How can I provide you with evidence if you won't let me amass that evidence?!' So, anything that can provide a short-term win without entailing huge risk or investment of time or resource, and can therefore provide evidence for the potential success of the project or change, can be very helpful.

Consolidate Early Gains

Once you have achieved some short-term wins, you must consolidate these early gains by pushing ahead for more change. The Wham O Corporation, who enjoyed success with a number of fun toys such as the hula hoop and the frisbee, found its niche in cheap, usually plastic-moulded, toys for all ages. Its approach is to keep innovating, recognising that such toys are subject to crazes that quickly move on, and as a result has been going for 70 years.

A performer who has a hit record needs to get out on the road to consolidate their fan base with live performances, and follow up that hit with further hit material. The artist who has had a successful exhibition or whose picture has hit the headlines will need to keep painting and the novelist with a shortlist for a major prize to their name will be urged by their publisher for the sequel.

Anchor New Changes in the Culture

In organisations, this may mean ensuring that any practice or process that links back to the old culture is stopped. A new process or product or policy may be ignored if employees and customers alike can still go back to the old way of doing things.

When the Romans crossed a river to meet with one of their enemies, they destroyed the bridge behind them to cut off any chance of retreat. They reasoned that it offered a strong incentive to remain committed to the fight and fight harder! Perseverance is necessarily enhanced when giving up or admitting defeat is not an option.

Old processes and paperwork, policies, rules and cultural norms, as well as past successes, can all constrain new thinking and the implementation of change. Organisations must therefore make sure that they cannot be resuscitated.

Infrastructure and Resource to Facilitate Innovation

A major tension can arise when seeking to fund experimentation and development of new ideas, especially at times when budgets for 'business-as-usual' are also under pressure. Financial constraints can often pose challenges to creative work and hinder its progress, which can cause resentment.

The BBC has one of the world's finest natural history TV production teams; when senior management tried to introduce what they perceived to be a perfectly reasonable budgeting mechanism for the production process, conflict arose, and people threatened a walk out. As one senior producer said: "If I need to spend 6 months in the Russian tundra to get 30 seconds of film, then that is the cost and no fancy budgeting process will alter that!"

NESTA was established to facilitate innovation in a number of ways. It aims to find a new generation of radical creative practitioners working in the UK today. It asks for nominations of people and projects that are challenging the status quo and working determinedly for the benefit of their communities and beyond.

NESTA's research has shown how important physical clusters are in promoting innovation, particularly in the creative industries (23). The research suggests that a cluster provides an advantage by creating 'value chain linkages', harnessing shared infrastructure, facilitating knowledge spill-overs, and attracting skilled workers. To effectively promote innovation, these clusters require connectivity between contractors and collaborators of all kinds, firms, and customers.

NESTA concludes that innovation could be accelerated by 'catalysing latent clusters'—by identifying which sectors work well together and work synergistically—rather than trying to build new ones from scratch. This process can be enhanced by using universities and other technical institutes to support start-ups and other new ideas with research, and by engaging local neutral bodies such as chambers of commerce to build trust between organisations and thus facilitate wider collaboration.

We have already mentioned the value of multidisciplinary teams in Chapter 6, where diversity of background and experience helps to challenge narrow thinking and facilitate fresh insights. In the academic world there is a move to promote a 'culture of convergence,' whereby multidisciplinary teams from different institutions—commercial, governmental and academic—collaborate to develop new thinking and bring ideas to fruition.

In addition to different disciplines and experiences, there are certain individual characteristics that are important for members of multidisciplinary teams: an ability to communicate to those outside their respective discipline without being patronising or obtuse, a willingness to share information and cooperate even to the extent that they might miss out on the limelight themselves, and a commitment to finding a common language to define the problem.

This latter point may seem simple but can actually sometimes be problematic given the ever-more specialised terminology present in so many fields.

Commercialising

Turning an idea into a global brand requires investment, but investors are reluctant to spend their money when licences and patents are not secure. There is a sweet spot for investors, just beyond the proof of concept but before a fully commercial product is ready. A product is never totally ready before launch, but there is only so much that can be tested in the lab before rolling it out. The 'real world' is necessary to truly test a product.

An inventor needs investment to launch the product, and whilst this might represent a risk for an investor, the potential upside is considerable. Once a product is shown to be successful in the marketplace, subsequent investors may not be able to secure such favourable deals.

However, the experience of too many early failures might kill enthusiasm by backers to continue with their investment, thus condemning the idea before it is fully developed and reaches its full potential. So the point at which a product is tested in the marketplace must be given careful consideration.

For some inventors, the easiest way to commercialise an idea is to licence it to someone with more expertise. If the licensee fails to pursue the development of the idea however, there should be a mechanism for recovering the patent so that others can try. Langer Lab, for example, can make a licensee give up its licence if it doesn't use the technology.

There are a number of other mechanisms for innovations to gain the necessary financial support to launch. Crowdfunding has been a novel way to gain investment, by asking future customers to invest in the creation of the product. Tax credits and other subsidies have also been a significant factor in the growth of certain industries, such as the video game industry in Canada.

Venture capital houses are an important source of funding for start-ups and may take a different view of the project's risk profile, compared to a high street bank, meaning that they are more amenable to investment. Institutions like NESTA combine investment funding, support and research across a wide range of innovation. They have launched a Big Society Finance Fund to stimulate social investment. Finally, peer-to-peer lending services provide small businesses with access to loan lenders and capital sources.

Many new ventures start locally and gain a niche foothold in the market, but then how do they expand? NESTA champions 'mass localism' which aims to create the right conditions for such ventures to partner with local communities and co-create approaches that are individualised to that community. This is a variation of franchising.

6. Working with Resistance

There are many reasons why a good idea becomes derailed, such as a lack of economic benefit, or the threat of high implementation cost. An underlying factor that is ever-present but rarely acknowledged is known by the acronym WIFM—what's in it for me? As we discussed earlier, it is always important to address and explicitly demonstrate the benefits of any change or new idea for the key decision-makers and stakeholders. Otherwise, they may fail to support the change or ideas.

Some of the other reasons why resistance occurs are identified below.

Fear of Failure

A recurring theme in this book has been dealing with failure, and a feature of creative people is their capacity to prevent failure from inhibiting their motivation to succeed.

Edison had a third of patent applications rejected and most of his more than 1,000 patents did not produce useful products. Shakespeare and Beethoven also had variable quality in their output, but continued to produce nonetheless.

"There's a high price to pay for being creative—tireless work, solitude and isolation, failure and the risk of ridicule and rejection. The reality of creative work is that most artists will never sell their pieces, most actors and musicians will never make it big, most writers will never pen best sellers, most start-ups will end in failure, and most scientists will never make earth shattering discoveries. It's a price that most of us don't actually want to pay." (24)

The lesson for organisations and funders of innovation is to factor in time and resources for multiple experiments and to expect failures along the way. This, of course, is easier said than done, and early failure can be a huge source of friction between artist and backer. The speed with which many recording artists are dropped by their record labels if their first recordings do not succeed is testament to this.

For the innovator, whilst the conventional advice might be: 'be courageous, take risks and prepare for failure', there is a need to build a strong commitment from supporters to see them through these failures. Christopher Columbus received numerous rejections to his requests for support until King Ferdinand of Spain agreed to back his plan to sail to the Far East, even though he knew that it was a risky venture and that, according to many experts at the time, Columbus had grossly underestimated the distance he would have to travel to reach Asia going around the world in a westerly direction.

A number of recording artists, including rock legend Bruce Springsteen, have enjoyed significant success with their second or third album (or later), thanks to their label deciding to continue supporting them despite the lack of (commercial) success of initial recordings.

Sunk Cost

When 90% of the budget has been spent and the fatal flaw in the new invention emerges, what do you do? There is a natural tendency to press on in the hope that all will come right in the end. After all, since 90% has already been spent, what's another 10%? However, not only is this 10% better used elsewhere, there are also potentially overspends that may now arise in order to fix the problem.

It is, of course, embarrassing to admit that all that time and money spent has not yielded a positive result, and wilful blindness about the problem can emerge. Band-aid solutions are applied in order to salvage the time and money spent to date, yet the better approach would be to scrap everything and start again (or move on to a different project). Investors are often advised to cut their losses as 'the first loss is cheapest'.

In 1998, construction begun on a new footbridge over the River Thames in London, and in 2000, the Millennium Bridge finally opened. It was swiftly nicknamed the 'Wobbly Bridge' on account of the significant lateral sway that occurred, swinging the bridge from side to side as some 90,000 pedestrians crossed the bridge on its first day.

After just 2 days in operation, the bridge had to be shut for a further 2 years to fix this unfortunate fault, at a cost of a further £5 million. The potential for these oscillations had actually been identified during construction and were on the bridge's risk register, but no one had wanted to acknowledge and act upon them and admit an embarrassing design failure in doing so. In the end, the embarrassment and cost were even greater.

Wilful Blindness

Business failure is often associated with an inability to see what is happening in a business or a marketplace, and to make the necessary change—even to the extent of reinventing the business—fast enough to meet challenges end enjoy continued success.

The rise of disruptive business models in the accommodation and public transport industries has come from new entrepreneurs outside the respective businesses (Airbnb and Uber respectively), not from existing hotel chains or taxi companies that have reinvented themselves.

Wilful blindness occurs when we are stuck in the belief that there is only one approach, one way of doing business, or one solution to a problem. This is when we need to draw on idea generation techniques that can help us to challenge assumptions and beliefs, to wonder 'what if?' and to employ blue-sky thinking.

Masaru Ibuka, one of Sony's founders, grew tired of carrying around his cassette recorder—complete with internal speakers and a recording mechanism—whenever he travelled anywhere. So, he asked his team for a cassette player that was optimised for playback and headphone use. Thus, the Walkman was born.

Until this point, Sony's engineers had been locked into the belief that tape recorders required high-quality speakers and a recording capability. But the request for a cassette player that prioritised different functions forced the engineers to challenge the assumption that speaker quality and recording capability were the key functions, and they successfully created a product for a whole new market instead.

Personal Attachment

Passion, enthusiasm, commitment are all qualities that are seen time and again in innovators of all kinds. Their single mindedness is a key ingredient in their success, but it is also their downfall; innovators must beware of becoming too personally attached to their ideas as this can prevent any necessary evolution of their idea.

Force Field

(See Chapter 5 for tips on using force field analysis.)

The social psychologist Kurt Lewin once likened the pattern of our days to a stream running downhill. The water finds the easiest course to follow, and the outcome of our days often follows the path of least resistance, shaped by the various demands, obstacles and challenges that we face, just as the course of the water is shaped by the boulders in its path.

During a busy day, we might have many calls on our time and pressures to achieve certain tasks—deliver the report on time, pick up the children from school, meet a friend for dinner later on. These pressures are often in conflict, competing for our time and attention, and Lewin suggested they might be displayed as a force field.

When advocating for the adoption of a new idea, we need to consider what the forces acting on the decision-makers might be, and which will encourage or undermine their support. We can then consider what we might do to reduce the undermining forces and maximise the encouraging forces.

Working with Emotion

When there is resistance to a new idea, dialogue around the idea can often become heated, especially if the idea's originator is passionately committed to its implementation. When we become emotional about a particular idea, logic and facts invariably get distorted.

Work by neuroscientist David Rock (25) suggests that there is a similar brain response to social threat as there is to physical threat. In particular, threats to our sense of status, certainty, autonomy, relatedness and fairness (combined in the acronym SCARF) evoke the same sense of psychological danger as physical threats and pain. Given that new ideas can certainly challenge individuals (and organisations) in these areas, it's not surprising that they can lead to strong emotional responses and push back.

For example, new ideas can challenge an individual's status and feelings of competence. ('I should have thought of that'. 'It's my job to develop new products'.) New ideas can also disrupt existing patterns of working and create uncertainty as a result.

Teams may be broken up (threatening a sense of relatedness), workloads reassigned and processes realigned (threatening a sense of fairness), and some people may feel that they no longer have control and their freedom to choose and act accordingly are diminished (threatening their sense of autonomy).

Disruptive ideas may also divide people in new ways, between those who support the new system and those who resist (see the current state of western politics for a vivid example!). Therefore, existing patterns of relationship are broken, with extremely damaging psychological impact.

By being aware of how the SCARF model operates, and how the introduction of new ideas may greatly threaten people in these areas, organisations and individuals can better navigate the process of discussing and implementing change.

Chapter Summary and Conclusions

In this chapter, we identified the ingredients in bringing new ideas to successful implementation. These include the following considerations:

- Employ a persuasive approach to build a constituency of support, and confidence in the approach and idea.
- Commitment can release the effort and hard work often required for successful innovation.
- Combine this commitment with a demonstration of the practicalities of implementation to drive success.
- Change of all kinds is invariably associated with resistance from some quarters and success is only assured when this is overcome. Pursue an approach that does not leave a legacy of increased resistance in the future.

Chapter 9
Creative Societies and Cities, and a Look to the Future

Introduction

Throughout this book, we have explored creativity on an individual level, and examined its place within organisations. In our final chapter, we examine creativity on a broader level—within the cities, cultures and countries that we inhabit, and discuss the impact that external factors have on our efforts to be creative.

We will explain how ideas spread through communities and societies, and how those communities and societies foster or hinder creativity—through their cultivation of the creative class of workers, their educational systems and attitudes to intellectual property, and through national initiatives.

We will discuss some of the considerations in these areas that will shape our capacity to be creative in the future. And finally, we also look at some of the ways in which our understanding of creativity is developing as research in the field of neuroscience continues to make new discoveries.

External Factors of Creativity

Remember the three ingredients of creativity that we discussed in Chapter 2? Stanford professor Tina Seelig codified those three ingredients as knowledge, imagination and attitude. In her model of creativity—her Innovation Engine—she also included three other ingredients that had an impact on our capacity to be creative, but were external—resources, habitat and culture (1).

Let's explore each of those external factors now.

Resources

Resources include all the materials we have at our disposal with which to create. They might be tangible—parts and materials with which to construct a new gadget—and they might be intangible—time, money, information, access to infrastructure, the knowledge we can glean from other people's experiences.

For an artist, materials such as canvases and paints and inks would be important resources; without access to these materials, the artist's capacity to create would be compromised. For some artists, these materials may be cheap and abundant; for others, they may be expensive and scarce. This will impact their relative approaches to their creative work.

The discovery of the Higgs Boson particle—often known as the 'God particle' since it is the fundamental building block of the universe—required the construction of an extremely expensive and complex tool, the Hadron Collider, to discover it. Without this resource and the huge investment required to construct it, such a discovery would not have been possible.

It is important to recognise the resources we have at our disposal. Often, we can be blind to them, particularly the more intangible ones, or take them for granted (e.g., access to all the different perspectives and bodies of knowledge that we can access when we are university students). We all have different resources at our disposal; beware of believing that the 'grass is greener' elsewhere.

Explore your world around you to uncover the resources you do have. Do research into grants that might be available to support your creative work; look for local support groups for like-minded creators who could provide fresh perspectives; take advantage of the charity shops that sell materials to experiment cheaply.

The unique combination of resources at your disposal is the key to unlocking the unique and creative ideas you could come up with. The creation of the Black and Decker Lawnraker, discussed earlier, was a result of taking pre-existing items—resources that the inventor could access—and putting them together in a novel way.

Habitat

This sums up the extent and context of the physical space in which we operate. What are the pertinent features of that habitat that influence our capacity to be creative? The spaces in which we do our work (offices, studios, workshops, laboratories, apartments), the bustle of a cityscape or the serenity of rural isolation, the nature of people with whom we interact...these factors can all have an impact.

We can control some—but not all—of this, and what we can control can be powerful. Within an office environment, we might not choose whether we have a cubicle or an office or an open plan desk, but we can choose how to decorate our space, whether to work in a messy or tidy space, whether to get out into the world outside during our lunchbreak or not. Often, we are limited only by our own imagination in our capacity to influence our habitat and our capacity to create.

Throughout this book, we have discussed some of the ways that this immediate environment around us has an impact on our creativity. Consult Chapter 3 for considerations on the impact of features of our habitat on our individual creativity, and Chapter 6 for a discussion of the impact of organisational habitats.

Culture

We discussed culture in organisational terms in Chapter 6. Remember the example of different greetings as an illustration of how different cultures can take different approaches, and how we are motivated to 'get it right' and avoid 'getting it wrong'. The behaviours that our cultures encourage and discourage also have an impact on our capacity to be creative on an individual level, by controlling how likely we are to engage in the behaviours that can lead to creativity.

Imagine that you find yourself in a culture that reveres and celebrates its individual entrepreneurs (e.g., Elon Musk); then imagine that you find yourself in a culture whose business figureheads are the CEOs of the biggest corporations (e.g., Fujio Cho of Toyota). Which culture will be most likely to encourage creative output from its inhabitants?

The answer is of course very nuanced as both cultures will encourage certain behaviours that will be beneficial to creative output. But they will do so in different ways and in different areas of the creative process—encouraging people to pursue their own ideas, or encouraging people to seek to rise to the top of the biggest and most powerful organisations (that can then pour time and money into innovation). Creative outcomes may therefore be different according to the cultural values.

Our culture houses our collective attitudes to failure. How forgiving are we as a society and culture? Do we permit people to make mistakes, learn from them, and try again, or do we hold them forever accountable? This balance can have important implications for our capacity to strive for creative outcomes.

Without an appropriately failure-tolerant culture, it is unlikely that inventor James Dyson would have persevered over the course of 5,000 attempts to bring his revolutionary vacuum cleaner to life. Are we encouraged to voice our own opinions, or are we encouraged to keep our heads down and conform to the prevailing wisdom? Are we encouraged to travel the world, and in doing so, to broaden our palate of experiences and perspectives? Or does our culture place more value on staying at home and remaining in our own country?

All these cultural values will have a bearing on the extent to which we undertake the behaviours, actions, dispositions that we have discussed throughout this book, on an individual level, collectively within our social groups, and organisationally. And this will have an impact on our capacity to be creative.

That's not to say that culture necessarily prohibits creativity, but it can certainly shape the ways in which we approach creative work, and can provide a supporting force or a slowing force in certain areas of the process.

As Seelig sums up, "Your *knowledge* provides the fuel for your imagination. Your *imagination* is a catalyst for the transformation of knowledge into ideas. This process is influenced by a myriad of factors in your environment, including your *resources*, *habitat*, and *culture*. Your *attitude* is a powerful spark that sets the Innovation Engine in motion" (2).

How Ideas Spread

In July 2012, a Korean rapper called Psy released his new single, *Gangnam Style*, together with its accompanying music video. To this point in his career, he had enjoyed some success in South Korea with 5 previous albums, but was not known to Western audiences.

However, the video for *Gangnam Style* suddenly started getting attention, first in countries with close ties to South Korea, such as the Philippines, and then globally. People viewed and shared it. Celebrities tweeted about it, notably the rapper T. Pain in late July, pop singer Katy Perry in August and even the president of South Korea (3).

Swiftly, its viewing numbers began skyrocketing, making it the first video to gain over 1 billion views on YouTube by December. How did this happen? What caused the video to take off and go viral?

There are certain elements that appear to have contributed to the video's spread. Firstly, the song is extremely catchy, with a simple key lyrical catchphrase ("opa Gangnam Style") that transcends linguistic barriers. Secondly, the video is highly entertaining with numerous surprising, amusing and memorable elements—the comically serious enthusiasm of Psy's delivery, the iconic horsey dance, the numerous backdrops and cameo appearances by Korean celebrities (although this might not necessarily appeal to those outside of South Korea).

Once celebrities began tweeting about the video, it was exposed to a wide audience of their fans. The video also provoked the creation of numerous participatory content—from videos of people simply watching the video, to parody videos that fans created themselves. Researchers at the Eotvos University in Budapest subsequently demonstrated that the song had spread in waves just like a disease, except across digital connections rather than through in-person contact.

As a measure of its success, only one other video passed the billion-views threshold in the next 2 years (Justin Bieber's music video for the song *Baby*), and *Gangnam Style* subsequently became the first video to pass two billion views in mid-2014. Almost a decade after its release, it is still in the top ten most viewed videos on YouTube, closing in on 4 billion views of the official post.

Over the past 10-15 years of social media, and the platform YouTube in particular, a number of pieces of video content have gone viral, from *Double Rainbow* to Rebecca Black's *Friday*, seemingly bursting from nowhere to widespread public consciousness…and then swiftly disappearing again to be occasionally recalled for the sake of nostalgia.

So why do videos and other pieces of online content, such as memes, go viral? Likewise, how do certain trends, fashions and creative ideas spread through societies and cultures? Who and what are the catalysts, and how do these influencers make this happen? How do YOU find out about new ideas, fashions, trends? From whom? And what prompts YOU to decide to adopt them yourself?

The Key Ingredients

There are certain elements that are the catalyst for the spread of ideas and, in the above examples, online content. Journalist and author Malcolm Gladwell wrote about the process in his bestselling book *The Tipping Point* (4), noting that "ideas spread just like viruses do" (hence *viral* videos, 'going viral'). According to Gladwell, there are three key elements at work in the spread of any idea: 'the law of the few', 'the stickiness factor', and 'the power of context'.

The 'law of the few' stipulates that there are a few key individuals who are integral to the spread of an idea. These include the 'influencers' that have become a familiar part of our cultural landscape, not least through their inescapable presence on social media. These individuals fall into three categories: mavens, salesmen, connectors. The mavens are the experts who unearth, or create, or expose the idea; they have influence through their expertise.

The salesmen are those that can convey the idea in a very compelling and persuasive way; they have influence through their persuasion. And the connectors (described earlier as 'telephone exchanges') are those people who know a lot of others and have a lot of connections, and can spread the idea far and wide through their broad network; they have influence through the size and breadth of their network.

Celebrities often act as both salesmen and connectors (and sometimes as mavens too)—they are persuasive and compelling on account of their celebrity status and expertise in their particular field (if they have one), and they are also connectors on account of their sizable fanbase, often easily reachable as a significant online/social media following. The artist Beyoncé, for example, currently has over 140 million Instagram followers, and is amongst the top 10 most followed accounts.

Kevin Allocca, Head of Culture and Trends at YouTube, explained the phenomenon of the infamous *Double Rainbow* video that went viral in 2009 and pinpointed a critical moment in the video's spread—when TV personality Jimmy Kimmel shared the content on Twitter, having been alerted to it by a friend of his (5).

It can be easy to forget that there are multiple roles in the spread of an idea. Of course, the originator of the idea is all-important. But a leader needs followers. Ideas don't spread without the followers who are the salesmen and connectors who must accept and amplify the idea, and it is important to consider who these people might be in the spread of any idea.

There is a video online of a festival-goer dancing at the Sasquatch music festival in Canada. At the start of the video, he is dancing alone, and someone has decided to film him, perhaps as entertaining content to be shared. Here the tag of 'madman' that we referenced earlier may ring true—as he dances, people watch him with amusement. But then a couple of brave souls decide that it actually looks like fun to dance, and they join him. Maybe they are being ironic. Who knows?

However, their friends then join them too, and this makes it a little bit more acceptable for anyone else to join in. Little by little, more people start dancing with them, and then suddenly, the tipping point that Gladwell references is reached. There are now enough people dancing that no one will be singled out. People come rushing over to dance, and very quickly those that are still sitting are in the minority.

The second ingredient—the Stickiness Factor—is the extent to which the idea or piece of content is 'sticky'; i.e., draws you in, is memorable, unexpected, impactful, shocking, catchy, etc. What makes this idea stand out? Why does it capture your attention? In the case of *Gangnam Style* described above, it was the catchiness of the song and the surprising and quirky humour of the video that made it sticky.

Finally, the power of context relates to the fact that there are contextual and environmental factors that will assist (or hinder) the spread of an idea. The London riots of 2010 spread across several days of hot summer weather, but some commentators have pointed out that it abated when the weather broke and rain came. Of course, there were many complex factors involved in the riot's spread and demise, but the contextual influence of the weather may be notable here.

In a TED talk (6), Allocca highlights the power of communities of participation in the spread of video content on YouTube. This refers to the online context that allows people to participate by modifying the content or creating their own version, and grow communities around it. This user-generated content is therefore a key element in the spread of an idea.

These shareable pieces of content have become known as memes, and they can spread rapidly as people use content in a variety of different contexts, or customise the content themselves to create their own variations. A classic example is a piece of footage from the Second World War movie *Downfall*, in which Hitler (played by Bruno Ganz) berates his staff in an increasingly apoplectic rage. The original movie dialogue was in German and employed subtitles, which have subsequently been changed to comment in a humorous way on all manner of other situations, from political decisions to developments in popular culture to personal disputes.

Ideas spread because a few critical factors are present in their spread; this is how creative ideas become successful innovations that are adopted across societies and beyond.

The Creative Class and Creative Societies

Have you ever wondered why some cities seem to be hotbeds of innovation and ideas? Why places like New York or Berlin or Budapest are known for their arts and music scenes? Why there are creative hubs in certain areas of the world, such as Silicon Valley? Whilst we all have the capacity to be creative—and all exercise our creative capacity on a regular basis one way or another—creativity has become increasingly institutionalised in society within the professions and careers that have emerged. And some cities have benefited considerably from attracting these professions and the people who work in those professions.

Civilisation has progressed through a number of stages of wealth-creation, from an agricultural society that accumulated wealth from owning land, to an industrial society that accumulated wealth from owning factories and means of production, to a knowledge society that accumulated wealth from knowledge… to our current phase in the western world as an ideas society that accumulates wealth from ideas.

As a result, the creative industries and those who work in them—the creative class—have increased significantly in economic value. A recognisable creative class has always existed, even as ideas of what constitutes 'creativity' have evolved. From ancient artisans to court jesters, creativity has been an important component of certain jobs. In some areas of creative output—the arts for example—the creative class relied on wealthy patrons to survive. In others, creators were independently wealthy.

For example, Henry Fox Talbot, who was responsible for a number of innovations in the development of photography, was also lord of the manor. This wealth allowed him the freedom to experiment with photography and to pursue the ideas that struck him. However, we are now in an age where the creative class is the driving force for economic development in post-industrial cities and communities. And it now earns a wage based on its valued creative output.

Since the creative class drives economic growth and prosperity, it figures t

hat those communities that can attract and retain the creative class will prosper, whilst those that don't will stagnate. The concentration in clusters of talented and creatively productive people is a powerful driving force of growth and prosperity.

Researchers such as Richard Florida have looked extensively at the nature of the creative class, as well as the cities and communities where they prosper. Florida has estimated that a third of the US workforce are part of the creative class, spread across certain sub-categories.

According to Florida's research (7), there are certain key ingredients that cities or communities need in order to attract (and retain) the creative class. They are the following 3Ts: talent, tolerance, technology. Talent refers to a talented, skilled, educated population; not surprisingly, top universities and university towns tend to be clusters of creativity.

The knowledge and expertise at the heart of creativity is required. A hub with talent will attract others with talent, perpetuating the presence of the creative class. The Cambridge science park—created when the University of Cambridge partnered with investors and entrepreneurs to establish a dedicated area nearby for these businesses to work on commercialising university research in entrepreneurial ways—was an early example.

Similarly, tech companies flocked to Silicon Valley, south of San Francisco, where other entrepreneurs, together with investors, were to be found.

Tolerance refers to the acceptance of individual tastes, sub-cultures, and cultural and lifestyle choices and preferences (e.g., sexuality) in a 'live and let live' ethos. With a prevailing tolerant climate, people are encouraged to express diverse ideas and opinions safely, even if they challenge the status quo (as creative ideas often do).

Finally, a measure of technology, in the form of technological resources and infrastructure necessary to fuel an entrepreneurial culture, as well as a concentration of high technology companies operating, is a key driver in building the presence of a creative class.

Creative workers are looking for cultural, social, and technological climates in which they feel they can best be themselves. Florida argues that they value meritocracy, diversity, and individuality, and look for these characteristics when they relocate. The presence of a creative class within a community fosters an open, dynamic, personal and professional environment, which in turn attracts more creative people.

This process can therefore become a self-perpetuating phenomenon as aspiring creators in certain fields migrate to those areas where established creators in the field are already to be found. Musical hubs and 'scenes' can spring up around the world in this way (such as the Los Angeles rock scene and Chicago house music scene of the 1980s, and the Seattle grunge scene and London drum n bass scene of the 1990s), and musical genres and styles evolve as ideas and approaches bounce between the hubs.

The New York Times described the recent musical phenomenon of drill music, a subgenre of hip-hop, as "a hyper-local strain of hip-hop that started in Chicago, was tweaked by bedroom producers in the United Kingdom, before taking over Brooklyn. Now it's the soundtrack to a summer of unrest" (8).

A question for cities, societies and communities to consider is how to establish clusters of businesses and engines of innovation that encourage creativity in specific areas, such as the video games industry that is concentrated on East London. These clusters often include academic institutions, start-ups, venture capital firms and other players who come together to encourage innovation.

Co-working spaces, incubators for entrepreneurial ideas and ventures, technology hubs are all examples of initiatives to drive the concentration of technological and entrepreneurial knowledge and facilitate the relatively free flow of ideas. It is this concentration of knowledge and free flow of ideas that has made Silicon Valley so successful as an incubator for new ideas.

Finally, however, one caveat for cities and communities to consider is that while attracting the creative class can bring prosperity to a city, this prosperity can be concentrated in the hands of a few, causing societal imbalance and inequality. How societies balance prosperity with equality is, of course, a complex but vital consideration.

Considerations for Education

According to the UK National Advisory Committee on Creative and Cultural Education, creativity is a rigorous process, based on knowledge and skill, that 'flourishes under certain conditions and can be taught' (9). Insofar as creativity contains processes and skills that can be taught, learned and developed, therefore, the attitudes and approaches to education espoused by societies and their communities are important.

We must be mindful to integrate into our educational systems the factors that encourage creative learning. We must consider the extent to which we promote aspects of education as a key pathway to creativity. And we must strive to enable the life-long learning that will help to sharpen our creative capacity.

Many elements of our educational systems still reflect principles designed to prepare children for the demands of a bygone era, particularly those of the industrial age. Author and educator Sir Ken Robinson has been particularly outspoken about some of the failings of the traditional educational model when it comes to fostering creativity, from its use of standardised testing to its reliance on right or wrong answers (10).

Certainly, there are ways in which our educational systems have evolved significantly with the times, not least in the adoption of technology into the teaching process—as academic subject matter, and as a means to impart teaching and learning. However, there are approaches to education and areas of the educational system that could be redesigned to better foster the creative capacities of tomorrow's societies.

Without necessarily abandoning education in traditional and fundamental areas such as reading, writing, and arithmetic, we must consider how we might place attention on the goal of promoting creative thinking and engaging students in challenging their own ideas, as well as those of their classmates and teachers. We must consider where we can encourage divergent thinking, rather than relying on a right/wrong dialectic. And we must seek to appropriately balance the placing of value on both academic effort and academic achievement.

A recent webinar hosted by Harvard Business Publishing, and featuring Professor Vijay Govindarajan of Dartmouth College in New Hampshire, highlighted three key enduring pillars of a traditional educational system: the transfer of knowledge from experts to students; the co-creation of knowledge through dialogue, debate, and collaboration; and the development of capabilities that are resistant to artificial intelligence (and cannot as yet be replaced by AI), such as creativity, emotional intelligence, moral judgement (11).

As we explore these educational pillars, we find a series of intriguing questions for us to consider, particularly as they pertain to fostering creativity.

Transfer of Knowledge

There is ongoing debate about the right balance in curricula between depth and breadth in subject matter, and the point at which any specialisation should occur. We have discussed in earlier chapters the value of a wide hinterland of knowledge and experience that encourages the mixing of different ideas, knowledge domains and experiences to create fresh insights and creative outcomes, and this would highlight the value of breadth. Yet the growing complexity of our world means that in many areas, deep study is required to acquire sufficient knowledge.

We see creative outcomes when knowledge is broad, facilitating the capacity to make unexpected and novel combinations of existing ideas, and when knowledge is deep, facilitating an awareness of the various parameters of a situation and the tools to play with. It is important for educators to consider how to balance the two paradigms of breadth and depth, to best incubate the next generation of polymaths accordingly.

In addition, we have seen the emergence of two opposing forces in education at play. On the one hand, we have moved towards greater conformity with national curricula, standards, and testing, and on the other hand, we have moved towards the integration of greater variety in our educational offerings, through electives and other opportunities.

Some educational approaches have moved, for example, from studying biology, physics and geography as separate subjects, to an integrated approach towards current issues in the guise of environmental studies. Whilst the former approach may be necessary to ensure appropriate rigour and quality across all areas of the educational system, it is the latter that upholds the importance of play, experimentation, and interconnectivity, valuing quirkiness and nurturing individual and zany interests.

These are the factors that have helped to encourage individuality rather than conformity, and have empowered our creative capacities. Fostering mental agility and curiosity, and providing opportunities for wide experience through travel and exposure to different cultures, for example, are now more commonplace, thanks to the increased variety of educational offerings.

Much education has become increasingly vocational, that is, geared towards equipping students with specialised knowledge for specific professions and careers. But to what extent is this a useful approach? Are today's teachings relevant for the world that students will encounter when they leave school in 10 or 5 or even 2 years' time?

Specific skills and knowledge can go quickly out of date, and an equal focus on equipping students to think, question, explore, experiment, and employ many of the principles of creativity that we have discussed throughout this book is surely important. This points to the importance of artificial intelligence-resistance capabilities, as discussed further below.

As vast amounts of information become distilled and instantly accessible through online search engines, the depth of knowledge that can play a role in creative outcomes is at our fingertips. The challenge then is no longer to effectively access this information, but to critically appraise the quality of this information.

Given the opportunity for anyone to create and share content online, the capacity to understand the nuances of quality, veracity, and bias—the truth, whatever that may mean—of any piece of information has become an increasingly critical aspect of education. The critical thinking that has long been an attribute of Western education has been particularly prized by students from China, who have come in thousands to study at Western universities in recent years.

Furthermore, this access to limitless information calls into question the need for students to memorise and hold information in their heads. If we have easy access to information, why do we need to remember it? Yet without this knowledge, how can our subconscious work on forming unexpected new combinations of existing ideas?

Co-Creation of Knowledge

The co-creation of knowledge occurs through discourse, and the process of bringing people together to discuss and debate ideas. (In Chapters 3 and 7, we looked at how to hold a creative discourse.) Educational forums can provide a safe space for a variety of perspectives to be aired and explored, and this provides the opportunity not only to expand our individual knowledge bases and perspectives, but also to bring different ideas and insights together to create fresh insights and creative and innovative ideas and solutions.

One of the factors that has changed the face of education is the ubiquity of inter-connected devices. As increasing percentages of the student population have access to not just a computer, but also to a mobile phone in their pocket—

devices that can connect them instantly to their friends, teachers, and broader social networks—the nature of interaction and co-creation and transfer of knowledge changes.

As technology affects the learning experience, it is imperative that we continue to foster shared learning spaces and enable the co-creation of knowledge and innovation through discussion and debate.

Developing Artificial Intelligence-Resistant Capabilities

As technology continues to change the world, we must continue to consider what the artificial intelligence-resistant capabilities that we need to instil might be. And we must explore the extent to which we can integrate teachings in "emotional intelligence, creativity, judgement, persuasion, complex pattern recognition", the artificial intelligence-resistant capabilities identified in the Harvard Business Publishing webinar (12), into our teaching curriculum. But we must also learn to harness the capabilities of artificial intelligence to *enhance* our creativity at the same time.

Throughout this book, we have highlighted skills, processes and capabilities that are key to the creative process and may also be artificial intelligence-resistant, such as divergent thinking, curiosity, the capacity to combine ideas across different domains. We must make sure we identify these capabilities and place emphasis on nurturing them within our educational systems. Many do not fit neatly into the traditional subject disciplines of education, nor the hard skills of a vocational education.

They are soft skills, whose value and even existence has been questioned in the past. Yet they make a difference—to one's capacity to be creative, and to one's capacity to effectively survive and thrive in an age of artificial intelligence.

So the challenge for educators is to consider how to build a development of these skills, processes and habits into the curricula that they teach.

The Arts in Education

There has been a particular emphasis in recent years on STEM subjects (science, technology, engineering and maths), which are certainly important disciplines in our technological age. However, the place of the arts in education as a tool to facilitate creativity is another area of consideration for our educational systems moving forward.

The arts are recognised as crucial in the process of developing creative students, and are often core disciplines for students at young ages. However, these same subjects are either cut from curricula in the later stages of schooling and education or moved to the side into electives that are devalued. Business demands for 'traditional' core competencies, and the need to provide students with vocational outcomes (i.e., help them get jobs) put pressure on schools to reduce timetable allocations of creative endeavours.

One study in Victoria, the second most populous state in Australia, found that the number of creative subjects undertaken as electives in the final year is relatively small (12%). And socio-economic background remains a major influence on children's access to arts-based disciplines such as music.

Yet the arts are where students gain the vital experiences of interaction, group work, composition and improvisation, performance, and using creative faculties for specific outcomes. If we want to foster increasing creativity within our educational systems, we might need to find a way to place an increased value on the role of the arts within education.

Legal and IP Protection

When it comes to creativity, an important concern lies in our approaches to intellectual property, and the balance of connecting ideas versus protecting ideas. On the one hand, we want to facilitate the connecting of ideas, since it is often the unexpected combination of existing ideas that leads to creative outcomes. However, we also want to protect ideas so as to incentivise and reward the originators of those ideas. This creates a very uneasy tension.

As more countries become compliant with international copyright treaties, protection for intellectual property increases, and this protection may increasingly encroach on our capacity to use and connect existing ideas in new combinations. Consequently, the systems we have established to protect intellectual property and the originators of creative ideas are also the very systems that may be stifling future creativity.

Furthermore, when companies protect ideas through patents and copyrights, but do so not to develop them but to prevent others from developing them, creativity and innovation are under real threat.

What is the answer here? This is an impossible question to address definitively, but we must be aware that the balance is delicate. And we must pay careful consideration to the patent and copyright systems and other mechanisms to protect ideas that we develop and uphold. We live in a world in which value is derived not only from individual innovations but also through interaction between individuals, so we must consider how this interaction might be encouraged while the resulting value is fairly distributed.

The music world in particular has been affected by this uneasy tension, and numerous successful pop artists—from Ed Sheeran to Katy Perry—have been sued in recent years for alleged copyright infringement. This is perhaps not surprising since so much popular music uses the same, common, shared musical 'vocabulary'. The extent to which any individual can lay an ownership claim to certain elements of this vocabulary is the source of much controversy.

Two recent cases have certainly highlighted some of the dilemmas here. Firstly, the artists Pharrell Williams and Robin Thicke were sued by Marvin Gaye's estate for copying the 'feel' of the Marvin Gaye song *Gotta Give It Up* in their hit song *Blurred Lines*, a 'feel' that the artists conceded was their inspiration for the track. Much debate has ensued as to whether it is valid for a 'feel' to be copyrightable.

Without debating the specifics of musicology or copyright law, we might note that every new creation has its genesis and its inspiration in something else that came before, and that inspiration is often evident in what follows.

The second case centres on the famous Led Zeppelin song *Stairway to Heaven*. Led Zeppelin were sued by the estate of Randy California, guitarist and songwriter with the group Spirit, who were contemporaneous to Led Zeppelin during the time of *Stairway to Heaven*'s composition.

The debate has focussed around whether the elements of the song that were allegedly copied—a descending chromatic progression used in *Stairway to Heaven*'s introductory chord progression and through the verses—are actually part of the shared music language, used commonly in baroque music from hundreds of years previously.

This progression is certainly a common musical figure, so should it be available to both artists to use, even if one was inspired to use it after hearing the composition of the other?

Since artists are drawing upon a shared musical language, and there are well-established conventions and tastes in what audiences want to hear, commonalities between songs are inevitable. By way of illustration, Australian comedy trio Axis of Awesome perform a song, the frankly-titled *4 Chord Song*, which cycles through a large number of well-known songs that all use the same chord progression, highlighting the blatant similarities between them. So, at what point does any of this become copyrightable?

Other areas are similarly plagued by patent and copyright issues, from disputes in the pharmaceutical industry over the use of certain patents or the development of generic versions of patented drugs, to the ongoing battles and 'patent wars' between smartphone manufacturers over various innovations in their respective products.

There are no easy answers here, but it must be recognised that the balance between protecting ideas and connecting ideas is a delicate one, but one that can have significant ramifications for creative practice.

National Initiatives

While some may see the combination of creativity and government as an oxymoron, it is important to recognise that governments play a significant role in supporting creative societies. Their approaches to, and investments in, education, regulation, intellectual property, public broadcasting, research, funding and other mechanisms of economic support, are impactful. Unfortunately, many of these areas of activity are constantly under threat of reduced scope and funding cuts, yet their survival and ongoing development are critical to the support of creativity on a national level.

If creativity is so valuable to humanity at all levels of society, how do we ensure that its position is prioritised at a governmental level? Should there be a Secretary for Creativity, a cabinet level position responsible for the nation's creativity and creative output? Political leaders could turn here for radical, breakthrough solutions to difficult challenges.

How can governments play a role in encouraging interaction? The town square has always been a place for debate and discussion throughout history. As society evolves—and interaction moves increasingly online—what are some of the ways that governments can continue to provide and facilitate meeting places for the exchange of ideas?

Around the world, on a local and international level, countries and communities experiment with ideas, from the creation of DesignSingapore—a national agency set up by the Ministry of Communications and Information to support designers in Singapore, part of a mission to build a 'creative economy'—

to the City of Sydney's discussion papers, workshops and round table debates on creativity for Sydney in Australia.

Denmark has also made design and creativity a national priority through initiatives such as the INDEX project, an international design competition that focusses on using design to improve life and address the big problems that humanity faces. And in the UK, institutions such as the innovation foundation NESTA provide seed money for entrepreneurial ventures.

Governments can also stimulate innovation by drawing concerted attention to particular causes, and with the offer of reward. As we've previously mentioned, prizes have often been a stimulus for innovation, from the Ansari X Prize for space flight, which netted $10 million for the Tier One project, to the $25 million prize up for grabs for technology capable of cooling the planet's climate.

A study, conducted by management consulting firm McKinsey, of prizes worth more than $100,000 concluded that the aggregate value of such large awards has more than tripled over the past decade, now totalling $375 million (13).

It is also crucial that governments encourage collaboration—both within their communities, across diverse industries and disciplines that are crucial for creative thinking and action, but also internationally. In particular, global issues, such as climate change or health pandemics, require global solutions, and countries must collaborate across international borders to find these global solutions.

The recent Covid 19 pandemic has starkly illustrated this need for international collaboration—from sharing resources and medical supplies to collaborating on the search for a vaccine and its subsequent distribution; the attitudes and capabilities to do this start with government. World forums such as the World Economic Forum that meets annually in the Swiss town of Davos have created opportunities for major world challenges and problems to be discussed, and for fresh thinking to be aired.

Conversely, governments can stop the spread of innovation in its tracks by prohibiting its use, often for political reasons. TikTok, the Chinese-owned social media platform, has been buffeted by political pressure and threats of bans in India, the US and elsewhere. Another Chinese technology company (Huawei) that is involved in the development and implementation of 5G telecommunication technology has similarly run afoul of foreign governments and seen its operations blocked in certain countries.

Finally, governments must consider how they affect the prevailing cultural attitudes in ways that can support creativity. If our societies are to thrive, with creativity as their centrepiece, it is of utmost importance that our governments support this process.

Technology, Artificial Intelligence and Neuroscience

For a long time, it was assumed that whilst machines may take over the physical world of manufacturing and other areas of the supply chain, the arts, with its creative input, would remain solely the province of human creativity.

However, the rapid emergence of artificial intelligence, big data and machine learning has challenged this assumption. The author Yuval Noah Harari (14) discusses the case of EMI (Experiments in Machine Intelligence) where an audience could not tell the difference between a chorale written by Bach and one generated by EMI software. The program had analysed Bach's work and identified the underlying patterns and rules. Using this software, a researcher was able to 'compose' 5,000 chorales in the style of Bach in a single day. This software went on to imitate the styles of other classical composers with similar results.

An even more sophisticated program, called 'Annie', has developed these capabilities further, incorporating machine learning and new inputs from the external world. "Cope (the creator of EMI and Annie) has no idea what Annie is going to compose next." Are big data and artificial intelligence enhancing or replacing divergent thinking?

This might appear to merely be an imitation of existing composers, but with enough input, this kind of software could create all kinds of 'new' music and musical styles, combining the various inputs of previous work into new and unexpected combinations.

Recent technology has seen the development of artificial neural networks, or connectionist systems, computing systems that are vaguely inspired by the biological neural networks that constitute animal brains. Such systems 'learn' to perform tasks by considering examples, generally without being programmed with task-specific rules.

Advances in Neuroscience

Despite everything that we have discussed in this book, there is still much for us to learn about the processes and experiences of creativity in action. Research in neuroscience continues to give rise to new theories about the nature of creativity and the associated processes in the brain, though this does not always lead to clarity and consensus.

The rapid advances in neuroscience have produced some exciting insights into how the brain produces novel thoughts. Research has focussed on understanding how the brain generates the spontaneous and relatively unconstrained thoughts that are experienced when the mind wanders, during daydreaming and other unfocussed thought, some of which is related to the creative process, particularly the incubation period.

Research (15) suggests that a brain system called the default mode network (DMN) is at play here. The DMN is a series of regions of the brain that seem to be interconnected and are especially active when an individual is not focussed on the outside world, and when cognitive control is low (e.g., we are just daydreaming or our minds are wandering rather than paying attention to a specific external task).

Research shows that the DMN has been related to complex, evaluative and unconscious forms of information processing and allows cognition to be guided by information that is not part of our immediate perceptual input (i.e., not stuff that we are perceiving in the world around us). Instead, the hinterland of knowledge and experience, that contributes to the flourishing of creativity, is at play here.

A number of studies are starting to show that the DMN might be involved in creativity (16). One study found a negative correlation between creativity and activity in the lingual gyrus (an area of the brain linked to visual processing and logical analysis), and a positive correlation between creative performance and grey matter volume in the right brain area that is part of the DMN.

Another study found a correlation between differences in creative performance and difference in volume of DMN. In other words, a greater volume of grey matter in the DMN provides more neural resources for generating creative ideas. These findings suggest (and provide neurological support for the argument) that less controlled cognitive processes, such as mind-wandering and daydreaming, are important in creativity.

Studies on neuroimaging have yielded other theories that also have an impact on creativity. For example, research demonstrates that the lateral prefrontal cortex impairs problem-solving, and that insight is the culmination of a series of brain states and processes operating at different time scales (17). How these findings might be harnessed to help us increase and improve our creative outcomes is as yet unclear, however. Ultimately, as Dietrich and Kanso (18) concluded after reviewing 72 neuroimaging studies on creativity and insight, the neuroscientific literature on creativity thus far is self-contradictory; creative thinking does not appear to critically depend on any single mental process or brain region. The DMN can be considered one, but not the only, single neural underpinning of creativity.

The Future

As we move forward into the future, some of these societal and cultural factors are ripe for change. There are possibilities for enhancing the creative capacities of our societies, individually, organisationally, institutionally and nationally.

We started this book with historic stories of creativity and innovation that enriched nations and changed the world. However, recent commentary suggests that employees within organisations are pushing back against the constant demands for 'innovation'. It might therefore be necessary to frame such a focus in a different way. But whatever semantics we use, the issue of driving innovation remains of utmost importance, for we live now in an evolving world with rapid and constant technological change, as well as the need for ever-more innovative and sophisticated solutions to manage our planet.

Global connectivity (as well as the backlash against it) and the instantaneous flow of speech, images and data are swiftly transforming collaboration, and this evolution will continue (global politics permitting). The recent global pandemic of Covid 19 has spurred fresh impetus for global collaboration on the development of a vaccine and treatments.

And with job markets and traditional forms of labour increasingly challenged by technology and by economic forces, there are many questions of what people will do to earn a living. The accelerating pace of change places an ever-higher premium on not only the new idea but also the ability to bring it to market quickly, as its lifespan shortens. The history of previous labour-market revolutions (the agricultural and industrial revolutions) suggests that new jobs and careers will emerge.

However, there is no ironclad law for the future, and enhancing and stimulating our creativity may well be our best strategy for finding new ways of living in a better world.

Appendix

Uses for a Paper Clip

- Clip paper together
- Reset a mobile phone
- Link together to form a chain
- Create an electrical connection
- Remove hair from your hairbrush
- Hold flowers in place
- DIY bookmark
- Reset electrical items
- Open an envelope
- Unclog a salt shaker
- Clean your nails
- Dip an egg into boiling water
- Hold bank notes together (money clip)

Broken Squares Game

This exercise is a game. Each team player will be given an envelope containing some puzzle pieces. The purpose of the game is to 'win' by assembling 5 equal-sized squares—one in front of each team member. The game will be over after all team players have a square of equal size in front of them.

There are two rules: Firstly, you cannot *take* pieces from someone else, you can only *give* pieces. Secondly, you cannot talk.

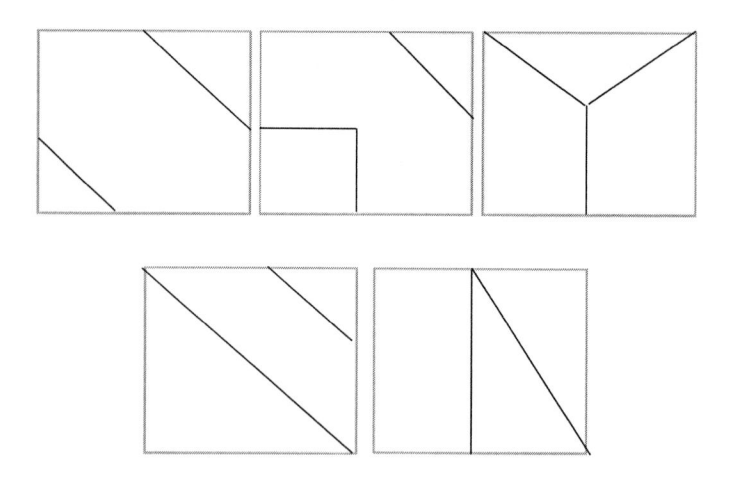

A Gantt Chart

A Gantt chart is a type of bar chart that is used in project management, named after its inventor, Henry Gantt. It can be used to show the timeframe for each activity in a project, the timing of each activity in relation to another activity, as well as how each activity fits into the overall timeframe. It can also be used to show the planned time for each activity against the actual time spent.

An example Gantt chart

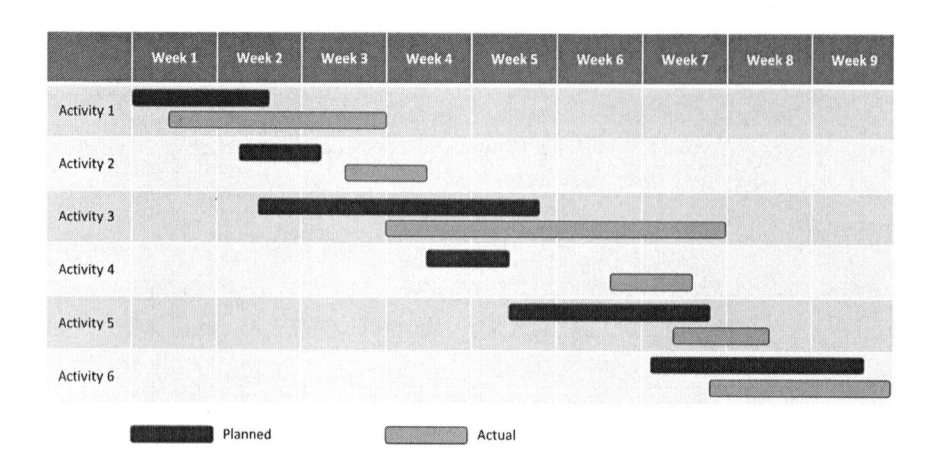

References and Bibliography

Introduction

1. Quint, J. (2016). *The Genius Way to Always Find Your Lost Earring*. PureWow. https://www.purewow.com/home/how-to-find-a-lost-earring
2. Altshuller, G. (2001). *40 Principles: TRIZ Keys to Technical Innovation* (L. Shulyak & S. Rodman, Trans.). Technical Innovation Centre.
3. IBM. (2010). *Global CEO Study: Creativity Selected as Most Crucial Factor for Future Success*. IBM. https://www.ibm.com/news/ca/en/2010/05/20/v384864m81427w34.html
4. IBM Institute for Business Value. (2012). *Leading Through Connections. Highlights of the Global Chief Executive Officer Study*. https://www.ibm.com/downloads/cas/3O8OG8RL
5. World Economic Forum. (2016). *Annual Report 2015-2016*. http://www.weforum.org/reports/annual-report-2015-2016
6. Florida, R. (2002). *The Rise of the Creative Class*, Basic Books. See also http://www.creativeclass.com/richard_florida
7. New York Times, 19 July 2010.
8. Taylor, C. W. (1959). The 1955 and 1957 research conferences: The identification of creative scientific talent. *American Psychologist, 14*(2), 100-102.
9. Sawyer, R. K. (2012). *Explaining Creativity: The Science of Innovation*. Oxford University Press, p. 11.
10. Johnson, S. (2011). *Where Good Ideas Come From*. Riverhead Books.
11. Dietrich, A. (2014). The myth conception of the mad genius. *Frontiers in Psychology, 5*, 79.
12. Sternberg, R. J., & Kaufman, J. C. (2010). *The International Handbook of Creativity*. Cambridge University Press.

13. Robinson, K. (2015). *Creative Schools. The Grassroots Revolution That's Transforming Education*. Viking Books. Includes references to the studies by Beth Jarman and George Land.

Chapter 1

1. Sawyer, R. K. (2012). *Explaining Creativity: The Science of Innovation*. Oxford University Press, p. 7.
2. Ferguson, K. (2015). *Everything is a Remix*. https://www.everythingisaremix.info
3. Lethem, J. as cited in Kleon, A. (2012). *Steal Like An Artist: 10 Things Nobody Told You About Being Creative*. Workman Publishing.
4. Sawyer, R. K. (2012). *Explaining Creativity: The Science of Innovation*. Oxford University Press, p. 8.
5. Sternberg, R. J., & Kaufman, J. C. (2010). *The International Handbook of Creativity*. Cambridge University Press.
6. Pascale, R. (2000). *Surfing the Edge of Chaos*. Currency.
7. Wallas, G. (1926). *The Art of Thought* (1949 ed.). Watts.
8. Dodds, R. A., Smith, S., & Ward, T. B. (2002). The use of environmental clues during incubation. *Creativity Research Journal, 14*(3&4), 287-304.
9. Sachs, O. (2017). *The River of Consciousness*. Knopf.
10. Watterson, B. (1990, May 20). *Kenyon College Commencement Speech*.
11. Quoted in Forbes magazine on 15 September 1974 as cited in Chang, L. (2006). *Wisdom for the Soul*. Gnosophia Publishers.
12. The Famous People. (n.d.). *26 Inspiring Quotes By Aaron Copland That Will Tug At Your Heartstrings*. https://quotes.thefamouspeople.com/aaron-copland-321.php
13. Sadler-Smith, E. (2015). Wallas' four-stage model of the creative process: more than meets the eye? *Creativity Research Journal, 27*(4), 342-352.

Chapter 2

1. Craft, A. (2000). *Creativity Across the Primary Curriculum: Framing and Developing Practice*. Sage.
2. Thorne, H. (2019). *Understanding Your Creativity*, Self-published.
3. Shapero, A. (1989). *Managing Professional People: Understanding Creative Performance*. Touchstone.
4. Currey, M. (2013). *Daily Rituals: How Artists Work*. Knopf.
5. Gregoire, C. (2014). *18 Things Highly Creative People Do Differently*. Huffington Post. https://www.huffingtonpost.com.au/entry/creativity-habits_n_4859769
6. Dyer, J., & Gregerson, H. (2011, September 27). Learn how to think different(ly). *Harvard Business Review*. https://hbr.org/2011/09/begin-to-think-differently
7. Amabile, T. (2012, April 26). *Componential Theory of Creativity*. Harvard Business School. https://www.hbs.edu/faculty/Publication%20Files/12-096.pdf
   Amabile, T. (1998, September-October). How to Kill Creativity. *Harvard Business Review*, pp. 76-87.
8. Seelig, T. (2014). *InGenius: A Crash Course on Creativity*. Harper Collins.
9. Amabile, T. (1998, September-October). How to Kill Creativity. *Harvard Business Review*, pp. 76-87.
10. Gladwell, M. (2011). *Outliers*, Back Bay Books.
11. Kilgour, M., & Koslow, S. (2009). Why and how do creative thinking techniques work? Trading off originality and appropriateness to make more creative advertising. *Journal of the Academy of Marketing Science, 37*(3), 298-309.
12. Rank, O. (1989). *Art and Artist: Creative Urge and Personality Development*. Norton and Co.
13. von Oech, R. (2008). *A Whack on the Side of the Head* (Special ed.). Grand Central Publishing.
14. Amabile, T. (1998, September-October). How to Kill Creativity. *Harvard Business Review*, pp. 76-87.
15. Amabile, T. (1998, September-October). How to Kill Creativity. *Harvard Business Review*, p. 79.

16. Duckworth, A. L., Peterson, C., Mathews, M. D., & Kelly, D. R. (2007). Grit: perseverance and passion for long term goals. *Journal of Personality and Social Psychology, 92*(6), 1087-1101.
17. Csikszentmihalyi, M. (2013). *The Psychology of Discovery and Invention and Flow*. Harper.
18. Sawyer R. K. (2012). *Explaining Creativity: The Science of Innovation*, Oxford University Press, pp. 408 & 417.

Chapter 3

1. New Adele album based on drunken diary entries. (2012, February 28). *The Irish Examiner*. https://www.irishexaminer.com/arid-30541552.htmlx
2. Sennett, S., & Groth, S. (Eds.). (2011). *Off the Record: 25 Years of Music Street Press*. University of Queensland Press, p. 255.
3. Kalnejais, R. (2012, March). How I Write. *Time Out*, p. 71.
4. King, S. (2010). *On Writing: A Memoir of the Craft*. Hodder & Stoughton.
5. In a commercial for X Stationery at Officeworks.
6. Mehta, R., Zhu, R., & Cheema, A. (2012). Is noise always bad? Exploring the effects of ambient noise on creative cognition. *Journal of Consumer Research, 39*(40), 784-799.
7. Bond, K. (2012). *Music can boost your creativity*. Fine Art Views. https://fineartviews.com/blog/51975/music-can-boost-creativity
8. Wong, M. (2014). *Stanford study finds walking improves creativity*. Stanford News. https://news.stanford.edu/2014/04/24/walking-vs-sitting-042414/
9. Brandt, A., & Eagleman, D. (2017). *The Runaway Species*. Catapult.
10. *Data Visualisation Reinterpreted by VISUALISED*. (2012). Visualised. https://vimeo.com/visualised
11. Kashdan, T. B. (2018, January 26). The Mental Benefits of Vacationing Somewhere New. *Harvard Business Review*. https://hbr.org/2018/01/the-mental-benefits-of-vacationing-somewhere-new
12. *The Creative Brain: How Insight Works*. (2014). BBC Horizon. https://www.dailymotion.com/video/x1gn21d

13. White, P. (2020, July 1). How to build a stronger memory. *Harvard Business Review*. https://hbr.org/2020/06/how-to-build-a-stronger-memory
14. Weingarten, G. (2007, April 8). Pearls before breakfast: Can one of the nation's great musicians cut through the fog of a DC rush hour? Let's find out. *The Washington Post*. https://www.washingtonpost.com/lifestyle/magazine/pearls-before-breakfast-can-one-of-the-nations-great-musicians-cut-through-the-fog-of-a-dc-rush-hour-lets-find-out/2014/09/23/8a6d46da-4331-11e4-b47c-f5889e061e5f_story.html
15. *Ricky Gervais Tells a Story About How He Learned to Write*. (2013). Fast Company. https://www.fastcompany.com/3016916/ricky-gervais-tells-a-story-about-how-he-learned-to-write
16. Honey, P., & Mumford, A. (2000). *The Learning Styles Helper's Guide*. Peter Honey Publications.
17. Rozin, P., & Royzman, E. (2001). Negativity bias, negativity dominance and contagion. *Personality and Social Psychology Review*, *5*(4), 296-320.
18. See www.chindogu.com for examples of chindogu.

Chapter 4

1. Amabile, T. (1998, September-October). How to Kill Creativity. *Harvard Business Review*, pp. 76-87.

Seelig, T. (2014). *InGenius: A Crash Course on Creativity*. Harper Collins.
2. Berlinger, J., & Milner, G. (2005). *Metallica: This Monster Lives*. St Martin's Griffin.
3. Konnikova, M. (2016, March 11). How to beat writer's block. *The New Yorker*. https://www.newyorker.com/science/maria-konnikova/how-to-beat-writers-block
4. Cameron, J. (1992). *The Artist's Way*. Tarcher/Perigee.
5. Oshin, M. (2017, August 16). *The daily routine of 20 famous writers (and how you can use them to succeed)*. Medium. https://medium.com/the-mission/the-daily-routine-of-20-famous-writers-and-how-you-can-use-them-to-succeed-1603f52fbb77

6. Fairweather, D. (n.d.). *The Creative Process*. Toothpaste For Dinner. http://toothpastefordinner.com/
7. Pressfield, S. (2012). *The War of Art*. Black Irish Entertainment.
8. Duckworth, A. L., Peterson, C., Mathews, M. D., & Kelly, D. R. (2007). Grit: perseverance and passion for long term goals. *Journal of Personality and Social Psychology, 92*(6), 1087-1101.
9. Pink, D. (2009). *Drive*. Riverhead Books.
10. Pemberton, C. (2015). *Resilience coaching, a practical guide for coaches*. McGraw Hill Education. (See the Individual Resilience Questionnaire)
11. Langer, E. J. (2006). *On becoming an artist*. Ballantine Books.
12. Jeffers, S. (2006). *Feel the fear and do it anyway*. Ballantine Books.
13. Farson, R. E. (1963). Praise reappraised. *Harvard Business Review, 41*(5), 61-66.
14. Locke, E. A., & Latham, G. P. (2002). Building a practically useful theory of goal setting and task motivation. *American Psychologist, 57*(9), 705-717.

    Locke, E. A. (1996). Motivation through conscious goal setting. *Applied & Preventive Psychology, 5*(2), 117-124.
15. Carver, C. S., & Scheier, M. F. (1998). *On the Self-Regulation of Behaviour*. Cambridge University Press.
16. Brunstein, J. C., Schultheiss, O. C., & Maier, G. W. (1999). The pursuit of personal goals: a motivational approach to well-being and life adjustment. In J. Brandtstädter & R. M. Lerner (Eds.), *Action and self-development: Theory and research through the life span* (pp. 169-196). Sage.
17. Sheldon, K. M., & Elliot, A. J. (1999). Goal striving, need satisfaction, and longitudinal well-being: The self-concordance model. *Journal of Personality and Social Psychology, 76*(3), 482-497.
18. Church, A. H., & Silzer, R. (2009). The potential for potential. *Industrial and Organisational Psychology, 2*(4), 446-452.

Chapter 5

1. Milne, A. A. (1926). *Winnie the Pooh*. Methuen.
2. Rawlinson, J. G. (1971). *Creative thinking and brainstorming*. Routledge.
3. Kilgour, M., & Koslow, S. (2009). Why and how do creative thinking techniques work? Trading off originality and appropriateness to make more creative advertising. *Journal of the Academy of Marketing Science, 37*(3), 298-309.
4. De Bono, E. (1992). *Serious Creativity – Using the Power of Lateral Thinking to Create New Ideas*. Harper Collins.
5. Rawlinson, J. G. (1971). *Creative thinking and brainstorming*. Routledge.
6. Ward, T. (2004). Cognition, creativity and entrepreneurship. *Journal of Business Venturing, 19*(2), 173-188.
7. Kilgour, M., & Koslow, S. (2009). Why and how do creative thinking techniques work? Trading off originality and appropriateness to make more creative advertising. *Journal of the Academy of Marketing Science, 37*(3), 298-309.
8. McFadzean, E. (2000). What can we learn from creative people? The story of Brian Eno. *Management Decision, 38*(1), 51-56.
9. Kipling, R. (1902). The Elephant's Child. In *Just So Stories*. Macmillan.
10. Buzan, A. (1996). *The Mind Map Book: How to Use Radiant Thinking to Maximise Your Brain's Untapped Potential*. Plume.
11. Ishikawa, K. (1976). *Guide to Quality Control*. JUSE.
12. Drury, F. (n.d.). *Visual success map*. Go Make It Yours. https://www.gomakeityours.com/
13. Haven, K. F. (1994). *Marvels of Science: 50 Fascinating 5-Minute Reads*. Littleton Colo Libraries Limited.
14. Altshuller, G. (2001). *40 Principles: TRIZ Keys to Technical Innovation* (L. Shulyak & S. Rodman, Trans.). Technical Innovation Centre.
15. von Oech, R. (2008). *A Whack on the Side of the Head* (Special ed.). Grand Central Publishing.
16. Ritter, S. M., & Dijksterhuis, A. (2014). Creativity – the unconscious foundations of the incubation period. *Frontiers in Human Neuroscience, 8*, 215.

https://www.frontiersin.org/articles/10.3389/fnhum.2014.00215/full
17. Peters. T. (1992). *Beyond Close to the Customer*. Tom Peters. https://tompeters.com/columns/beyond-close-to-the-customer/

Chapter 6

1. IBM Institute for Business Value. (2012). *Leading Through Connections. Highlights of the Global Chief Executive Officer Study*. https://www.ibm.com/downloads/cas/3O8OG8RL
2. Bible (New International Version). Exodus 18:13-27.
3. Halliday, S. (2005, February 21). *Relentless Toyota thrives on chaos*. Adage. https://adage.com/article/toyota-report/relentless-toyota-thrives-crisis/102133#
4. Merriam-Webster Dictionary.
5. Ekvall, G. (1996). The organisational climate for creativity and innovation. *European Journal of Work and Organisational Psychology*. 5(1), 105-123.
6. Clegg, C., Unsworth, K., Epitropaki, K., & Parker, G. (2002). Implicating trust in the innovation process. *Journal of Occupational and Organisational Psychology 75*(4), 409-442.
See also Clegg, B., & Birch, P. (2006). *Instant Creativity*. Kogan Page.
7. Catmull, E. (2014). *Creativity Inc*. Random House.
8. Zhou, J., & Shalley, C. E. (2007). *Handbook of Organisational Creativity*. Psychology Press.
9. Christensen, C., & Eyring, H. J. (2011). *The Innovative University: Changing the DNA of Higher Education*. Wiley.
10. Heifetz, R., & Laurie, D. L. (2001, December). The work of leadership. *Harvard Business Review*. https://hbr.org/2001/12/the-work-of-leadership
11. Catmull, E. (2014). *Creativity Inc*. Random House.
12. Johnson, S. (2010). *Where good ideas come from*. TED. https://www.ted.com/talks/steven_johnson_where_good_ideas_come_from?language=en
13. Prokesch, S. (2017). The Edison of medicine. *Harvard Business Review*. https://hbr.org/2017/03/the-edison-of-medicine

14. Ritter, D. & Brassard, M. (2005). *The Creativity Tools Memory Jogger.* Goal/QPC.
15. Schumpeter: Diversity Fatigue. (2016, February 13). *The Economist*, p. 63.
16. Edison, T. A. (2003). *The Edison and Ford Quote Book.* Applewood Books.
17. Taylor, C., & LaBarre, P. (2007). *Mavericks at Work: Why the Most Original Minds in Business Win.* Harper Collins.
18. Tett, G. (2015). *The Silo Effect, the Peril of Expertise and the Promise of Breaking Down Barriers.* Simon & Schuster.
19. Sinek, S. (2014). Why we still treat our staff like line items. *Work*, p. 61.
20. Asimov, I. (2014). *How do people get new ideas?* MIT Technology Review. https://www.technologyreview.com/2014/10/20/169899/isaac-asimov-asks-how-do-people-get-new-ideas/
21. Schumpeter: Diversity Fatigue. (2016, February 13). *The Economist*, p. 63.
22. Pink, D. (2009). *Drive*, Riverhead Books.
23. Groves, K., & Knight, W. (2010). *I Wish I Worked There! A Look Inside the Most Creative Spaces in Business.* Wiley.
24. Kao, J. (1997). *Jamming.* Harper Collins Business.
25. Wong, M. (2014). *Stanford study finds walking improves creativity.* Stanford News. https://news.stanford.edu/2014/04/24/walking-vs-sitting-042414/
26. Coyle, D. (2018). *The Culture Code.* Random House.

Chapter 7

1. Loveless, A. M. (2001). *Literature review in creativity, new technologies and learning.* Futurelab. https://www.researchgate.net/publication/32231354_Literature_Review_in_Creativity_New_Technologies_and_Learning
2. Lane, M. (2010, October 14). *What's an egg race got to do with inventing?* BBC. https://www.bbc.com/news/magazine-11438219
3. Honey, P., & Mumford, A. (2000). *The Learning Styles Helper's Guide.* Peter Honey Publications.
4. Belbin. (n.d.). *The Nine Belbin Team Roles.*

http://www.belbin.com/about/belbin-team-roles/

5. Osborn, A. (2001). *Applied Imagination: Principles and Procedures of Creative Problem Solving* (3rd revised ed.). Creative Education Foundation Press.
6. Basadur, M. (1995). *Flight to Creativity: How to Dramatically Improve Your Creative Performance*. Applied Creativity Press.
7. Basadur, M. (1995). *The Power of Innovation: How to make innovation a way of life and put creative solutions to work*. Pitman Professional Publishing/Applied Creativity Press.

Chapter 8

1. Biography.com (2014). *Levi Strauss*. https://www.biography.com/fashion-designer/levi-strauss
2. No fear of flying. (2006, November 18). *The Economist*. https://www.economist.com/books-and-arts/2006/11/16/no-fear-of-flying
3. *Muhammad Yunus: 'Put poverty in the museum'*. (2013). DW. https://www.dw.com/en/muhammad-yunus-put-poverty-in-the-museum/a-16778589
4. Prokesh, S. (2017, March-April). The Edison of Medicine. *Harvard Business Review*, p. 134-143. https://hbr.org/2017/03/the-edison-of-medicine
5. Allan, D., Kingdon, M., Murrin, K., & Rudkin, D. (2002). *Sticky Wisdom: How to Start a Creative Revolution at Work*. Capstone.
6. Ludwig, A. M. (1995). *The Price of Greatness: Resolving the Creativity and Madness Controversy*. Guildford Press.
Ludwig, A. M. (1996). A Template for Greatness. *R&D Innovator*, 5(2).
7. Kaufman, S. B., & Gregoire, C. (2017). *Wired to Create: Unravelling the Mysteries of the Creative Mind*. Vermilion.
8. Machiavelli, N. (2003). *The Prince* (Reissue ed.). Penguin Classics.
9. Carnegie, D. (1936). *How to Win Friends and Influence People* (100th ed.). Simon & Schuster.
10. Fisher, R., & Ury, W. (1981). *Getting to Yes*. Penguin.
11. Covey, S. (1989). *Seven Habits of Highly Successful People*. Free Press.
12. Thaler, R., & Sunstein, C. (2008). *Nudge*. Penguin.

13. Lynch, D., & Kordis, P. L. (1988). *The Strategy of the Dolphin: Scoring A Win in a Chaotic World*. William Morrow and Company.
14. Kahneman, D., & Tversky, A. (2000). *Choices, Values and Frames*. Cambridge University Press.
15. Wintle, W. D. (n.d.). *Thinking*.
16. Dickens, C. (2004). *David Copperfield* (Revised ed.). Penguin Classics.
17. Ilczynski, M. (2017). *The Deal*. Seek.com.
18. Boccioni, U., Carra, C., Russolo, L., Balla, G., & Severini, G. (1910). *Technical manifesto of futurist painting*.
19. Ludwig, A. M. (1996). A template for greatness. *R&D Innovator*, 5(2).
20. Kahneman, D., & Tversky, A. (2000). *Choices, Values and Frames*. Cambridge University Press.
21. Kotter, J. (2012). *Leading Change*. Harvard Business Review Press.
22. Westlake, S., & Bunt, L. (2010). *Schumpeter Comes to Whitehall*. NESTA. https://www.nesta.org.uk/report/schumpeter-comes-to-whitehall/
23. Kaufman, S. B., & Gregoire, C. (2017). *Wired to Create: Unravelling the Mysteries of the Creative Mind*. Vermilion.
24. Rock, D. (2009). *Your Brain at Work: Strategies for Overcoming Distraction, Regaining Focus and Working Smarter All Day Long*. Harper Business.

Chapter 9

1. Seelig, T. (2014). *InGenius: A Crash Course on Creativity*. Harper Collins.
2. Seelig, T. (2014). *InGenius: A Crash Course on Creativity*. Harper Collins.
3. The Hollywood Reporter. (2013, February 25) *PSY Brings 'Gangnam Style' to Korean Presidential Inauguration*. Billboard. https://www.billboard.com/articles/columns/k-town/1549866/psy-brings-gangnam-style-to-korean-presidential-inauguration
4. Gladwell, M. (2002). *The Tipping Point*. Back Bay Books.
5. Allocca, K. (2011). *Why videos go viral*. TED. https://www.ted.com/talks/kevin_allocca_why_videos_go_viral
6. Allocca, K. (2011). *Why videos go viral*. TED.

https://www.ted.com/talks/kevin_allocca_why_videos_go_viral
7. Florida, R. (2008). *Who's Your City?* Basic Books.
8. Coscarelli, J. (2020, June 30). Pop Smoke Took Brooklyn Drill Global. Fivio Foreign Is Carrying the Torch. *New York Times*. https://www.nytimes.com/2020/06/30/arts/music/fivio-foreign-big-drip-brooklyn-drill.html
9. Craft, A. (2000). *Creativity Across the Primary Curriculum: Framing and Developing Practice*. Sage.
10. Robinson, K. (2006). *Do schools kill creativity?* TED. https://www.ted.com/talks/sir_ken_robinson_do_schools_kill_creativity
11. *COVID-19's impact on the future of higher education: what university leaders should be thinking now*. (2020). Harvard Business Publishing. http://academic.hbsp.harvard.edu/webinar_COVID-19_impact_on_the_future_of_higher_education
12. *COVID-19's impact on the future of higher education: what university leaders should be thinking now*. (2020). Harvard Business Publishing. http://academic.hbsp.harvard.edu/webinar_COVID-19_impact_on_the_future_of_higher_education
13. Bays, J., Goland, T., & Newsum, J. (2009). *Using prizes to spur innovation*. McKinsey & Company. https://www.mckinsey.com/business-functions/strategy-and-corporate-finance/our-insights/using-prizes-to-spur-innovation#
14. Harari, Y. N. (2016). *Homo Deus*. Vintage.
15. Ritter, S. M., & Dijksterhuis, A. (2014). Creativity – the unconscious foundations of the incubation period. *Frontiers in Human Neuroscience, 8*, 215. https://www.frontiersin.org/articles/10.3389/fnhum.2014.00215/full
16. Manni, R. (2005). Rapid eye movement sleep, non-rapid eye movement sleep, dreams and hallucinations. *Current Psychiatry Reports. 7*(3), 196-200.
17. Stickgold, R., & Walker, M. P. (2004). To sleep, perchance to gain creative insight? *Trends in Cognitive Science, 8*(5), 191-192.
18. Dietrich, A., & Kanso, R. (2010). A review of EEG, ERP and neuro imaging studies of creativity and insight. *Psychology Bulletin. 136*(5), 822-848.

Additional bibliography

- Andreasen, N. (1995). *The Creative Brain: The Science of Genius*. Dana Press.
- Barez Brown, C. (2006). *How to Have Kickass Ideas*. Harper Element.
- Berns, G. (2010). *Iconoclast: a neuroscientist reveals how to think differently*. Harvard Business Review Press.
- Buchsbaum, D., Gopnik, A., Griffiths, T. L., & Shafto, P. (2011). Children's imitation of causal action sequences is influenced by statistical and pedagogical evidence. *Cognition*, *120*(3), 331-340.
- Burrus, D. (2015). *Creativity and Innovation: your keys to a successful organisation*. Huffington Post. https://www.huffpost.com/entry/creativity-and-innovation_b_4149993
- Caird, S. (1991). The enterprising tendency of occupational groups. *International small business journal*, *9*(4), 75-81.
- Cannon J. (2003). *Moving into self-employment* [Unpublished doctoral thesis]. London University.
- Cannon, J. (2002). Changing me – changing you. The art of changing behaviour. In *City HR Review Directory* (pp. 7-12). Gee.
- Cannon, J., McGee, R., & Stanford, N. (2010). *Organisation design and capability building*. CIPD.
- Christensen, C., Dyer, J., & Gregerson, M. (2011). *The Innovator's DNA: Mastering the 5 skills of disruptive innovators*. Harvard Business Review Press.
- Colzato, L. S., Ozturk, A., & Hommel, B. (2012). Meditate to create: The impact of focussed attention and open monitoring training on convergent and divergent thinking. *Frontiers in Psychology*, *3*, 116.
- Crabtree, J., & Crabtree, J. (2011). *Living with a Creative Mind*. Zebra Collective.
- Davis, G. (1998). *Creativity is forever*. Kendall Hunt.
- De Dreu, C. K., Nijstad, B. A., Bechtoldt, M. N., & Baas, M. (2011). Group creativity and innovation: a motivated information processing perspective. *Psychology of Aesthetics, Creativity and the Arts*, *5*(1), 81-89.

- De Dreu, C. K. (2006). When too much and too little hurts: Evidence for a curvilinear relationship between task conflict and innovation in teams *Journal of Management, 32,* 83-107.
- De Sousa, F. C., Monteiro, I. P., & Pellissier, R. (2012). Creativity, innovation in organisations. *The International Journal of Organisational Innovation, 26,* 29-41.
- Drucker, P. F. (2002, August). The discipline of innovation. *Harvard Business Review.* https://hbr.org/2002/08/the-discipline-of-innovation
- Fredrickson, B. (2009). *Positivity.* Harmony.
- Furnham, A. (2016). *Appetite for risk: a sideways view.* Psychology Today. https://www.psychologytoday.com/blog/sideways-view/201609/appetite-risk
- Gray, A. (2016). *The 10 skills you need to thrive in the fourth industrial revolution.* World Economic Forum. www.weforum.org/agenda/2016/01/the-10-skills-you-need-to-thrive-in-the-fourth-industrial-revolution
- Grosz, S. (2013). *The Examined Life.* Chatto and Windus.
- Hunter, S. T., Friedrich, T. L., Bedell-Avers, K. E., & Mumford, M. D. (2007). Creative cognition in the workplace: an applied perspective. In M. J. Epstein, T. Davila & R. Shelton (Eds.), *The creative enterprise—managing innovative organisations and people* (pp. 171-193). Praeger.
- Hunter, S. T., Bedell-Avers, K. E., & Mumford, M. D. (2007). Climate for creativity: a quantitative review. *Creativity Research Journal, 19*(1), 69-90.
- Jamison, K. R. (1995). *Touched by Fire.* Free Press. Quoted in Carter, R. (2004). *Mapping the Mind.* Phoenix, p. 165.
- Jung, R. E., Segall, J. M., Bockholt, H. J., Flores, R., Smith, S. M., Chavez, R. S., & Haier, R. J. (2010). Neuroanatomy of creativity. *Human Brain Mapping, 31*(3), 398-409.
- Kandel, E. (2012). *The Age of Insight: The Quest to Understand the Unconscious in Art, Mind and Brain, from Vienna 1900 to the Present.* Random House.
- Kanter, R. M. (1989). *When Giants Learn to Dance.* Simon & Schuster.
- Katzenbach, J. R., & Smith, D. K. (2006). *The Wisdom of Teams: Creating the High-Performance Organisation.* Harper Business.

- Kim, Y. J., & Zhong, C-B. (2017). Ideas rise from chaos: information structure and creativity. *Organisational Behaviour and Human Decision Processes, 138*(January), 15-27.
- Konishi, M., McLaren, D. G., Engen, H., & Smallwood, J. (2015). Shaped by the past: the default mode network supports cognition that is independent of immediate perceptual input. *PLOS ONE, 10*(6), e0132209.
- Kühn, S., Ritter, S. M., Müller, B. C. N., van Baaren, R. B., Brass, M., & Dijksterhuis, A. (2013). The importance of the Default Mode Network in creativity—a structural MRI study. *Journal of Creative Behaviour, 48*(2), 152-163.
- There are many references to team building. See, for example, Lencioni, P. (2002). *The Five Dysfunctions of a Team: A Leadership Fable.* Jossey-Bass.
- Maccoby, M. (1991, December). The innovative mind at work. *IEEE Spectrum*, pp. 23-35.
- Majchrzak, A., More, P. H. B., & Faraj, S. (2011). Transcending knowledge differences in cross-functional teams. *Organisation Science, 23*(4), 907-1211.
- Marshall, A. (2013). *There's a critical difference between creativity and innovation.* Business Insider. https://www.businessinsider.com.au/difference-between-creativity-and-innovation-2013-4
- McNerney, S. (2012). *Unconscious creativity: step back to step forward.* Big Think. https://bigthink.com/videos/universal-basic-income-unemployment
- McWhinney, W. (1993). All creative people are not alike. *Creativity and Innovation Management, 2*(1), 3-16.
- Morley, E., & Silver, A. (1977, March). A film directors' approach to managing creativity. *Harvard Business Review.* https://hbr.org/1977/03/a-film-directors-approach-to-managing-creativity
- Negus, K. (2004). *Creativity, Communication and Cultural Value.* Sage.
- Paul, E. S., & Kaufman, S. B. (2014). *The Philosophy of Creativity: New Essays.* Oxford University Press.
- Robson, M. (1993). *Problem Solving in Groups.* Gower.

- Rogers, C. (1960). *On Becoming a Person*. Constable.
- Smallwood, J., Nind, L., & O'Connor, R. C. (2009). \When is your head at? An exploration of the factors associated with the temporal focus of the wandering mind. *Conscious Cognition*, *18*(1), 118-25.
- Smith, B., & Blagrove, M. (2015). Lucid dreaming frequency and alarm clock snooze button use. *Dreaming*, *25*(4), 291-299.
- Sun, R., & P. Fleischer. (2012). A cognitive social simulation of tribal survival strategies: The importance of cognitive and motivational factors. *Journal of Cognition and Culture*, *12*(3-4), 287-321.
- Sun, H., Wong, S. Y., Zhao, Y., & Yam, R. (2012). A systematic model for assessing innovation competence of Hong Kong/China manufacturing companies: a case study. *Journal of Engineering and Technology Management*, *29*(4), 546-565.
- Sun, R., & Helie, S. (2010). Incubation, insight, and creative problem solving: a unified theory and a connectionist model. *Psychological Review*, *117*(3), 994-1024.
- Tan, E. (2012). Special edition on innovation in organisations. *Learning Innovation Unit*. Dublin University.
- Tannenbaum, R., & Schmidt, W. (1958). How to choose a leadership pattern. *Harvard Business Review*, *36*(2), 95-101.
- Trickey, G. (2019). *Why your creative employees are more likely to be risk takers*. Training Zone. https://www.trainingzone.co.uk/develop/talent/why-your-creative-employees-are-more-likely-to-be-risk-takers
- Vangundy, A. B. (2007). *Getting to Innovation: How Asking the Right Questions Generates the Great Ideas Your Company Needs*. Amacom.
- Weick, K. E. (1995). Sensemaking in organisations. *Foundations for Organisational Science* (Vol. 3). Sage.
- Zander, B. and Zander, R. (2002) *The Art of Possibility: Transforming Professional and Personal Life*, Penguin Books.

Index

3
3M 181, 187

5
5 whys 77, 141, 146, 236, 252

9
9 Dots Exercise 117

A
Adele 62, 71
affinity diagram 126, 236
Affinity Diagram 127
Airbnb 284
Alcock 217
Allocca 292, 293
Altshuller 16, 158
Amabile, t 51
Amazon 72
anchoring 230
Angelou 42, 64
Ansari X Prize 265, 304
Apple 19, 36, 77, 177, 199, 266
Archimedes 15, 16, 22, 82
Asimov 207
Atlassian 187
attention 72, 74, 75, 76, 80
attitude 58, 61
Attribute Combinations 152
Auden 47
Auster 65
Axis of Awesome 302

B
Banksy 32, 33
Basadur Simplex Model 244
BBC 279
Beethoven 47, 59, 282
Beijing subway 30
Bell 75
Black and Decker 50, 70, 140, 164, 260, 288
Blanchard 268
Bohr 54, 128
Bond 68
Booker Prize 32
Boxes exercise 109
Boxes Exercise 111
brain gym 68
Brain Writing 126
brainstorming 121, 138
Branson 44, 78
British Airways 223
Britten 47, 68
Broken Squares Game 309
Brunelleschi 45
Bryant and May 77
Buckley 66
buddy system 202
burning platform 276, 277
Buzan 143

C
Callas 78
Cameron 89, 101
Canon 222

327

Carlson 258
Catmull 185, 222
Cavendish Partners 222
Change Management 275
Checklist for Innovative Teams 214
Chindogu 85
Cho 173
Christensen 179
Churchill 257
Classic FM 30
Clayderman 272
cognitive styles 233
Coleridge 65
Colossus 276
Columbus 283
Copland 39
Covey 263
Craft 44, 49
creative class 18, 294
creative hubs 294
creative people 44, 46, 47, 49
Critical Friend 102
Crowdfunding 281
Curie 42
Currey 47
Customer Ideas 163

D

da Vinci 54, 143
Daft Punk 56
Dali 157
Darwin 21, 22, 58
Davis 155
Day 170
de Bono 121
De Bono 115
default mode network 306
definitions of creativity 27
devil's advocate 115, 139, 184, 256, 274
Dickens 272
Dietrich 307
disruptive change 34
Dodds 38
Dragon's Den 275

Drawing the Problem 148
Duckworth 60, 92
Dyson 290

E

Earhart 78
Earle 135
Edison 21, 22, 59, 87, 190, 259, 269, 273
Einstein 27, 39, 44, 59, 77, 141
Ekvall 175, 205
Emanuel 276
empathic design 164
Empathic design 163
Eno 134
exposure 73, 80
Exposure 72
extrinsic motivation 59

F

Facebook 18
facilitation 228
fear of failure 89, 98
Feedback 268
Ferguson 30
Fishbone diagram 145
Fleming 157
Florida 18, 295
flow 60
Flow 61
force field analysis 169, 252, 285
Ford 21
Fox Talbot 294
Franklin 209
Fry 181

G

Gallagher 63
Gantt chart 254, 310
Gaye 302
Gervais 78
Gilbert 69, 101
Gladwell 52, 292, 293
Google 187, 222
Gore-Tex 202
Greene 88

Gregoire 48
grit 92, 93
Grit 60, 269
ground rules 231
Group Process 229
group think 182, 208, 256, 274
Gutenberg 71

H

Hamilton 16
Harari 305
Harley Davidson 164
Harrison 58
Harvard 15
Hazlett 270
Heathrow Express 217
Helfgott 53
Helpers and Hinderers 224
Hewlett Packard 173, 219, 223
Hickman 50
Higgs 288
high jump records 34
Hirst 31
Honey 80
Hopper 15, 16
Huawei 304

I

IBM 17, 170, 173, 218
idea generation techniques 106
idea management systems 240, 241
Ideas Matrix 153
illumination 40, 156
Imaginary brainstorming 125
Impact Analysis 166
implementation of ideas 176
improvement 34
improvisation 155
incubation 38, 306
innovation 33, 34, 36
Innovation Climate 174
Innovation Climate Questionnaire
 174
interpretation 72, 77, 79
Intimation 40

intrinsic motivation 88, 92, 107, 109, 186, 203, 218, 272
Intrinsic motivation 59, 61
Ive 19
Izzard 78

J

Jarrett 155
Jeffers 97
Jobs 18, 27, 42, 84, 222
Johnson 21, 186
Jordan 99
Joy Division 73
Juilliard School 53

K

Kahneman 273
Kaikaku 34
Kalnejais 63
Kao 222
Kekulé 156
Kellogg 96
Kipling 136
knowledge and expertise 51, 52, 53
Knowles 42
Kodak 198
Kolb 80
Kotter 275

L

Lady Gaga 29
Langer Lab 259, 260
Leadership Questionnaire 194
learning styles 80, 233
Led Zeppelin 302
Lethem 31
Lewis 135
Locke 100
Loveless 229
Lucas 30

M

Machiavelli 263
Madonna 29
Mahler 47
Martin 90

Maugham 47
McCartney 15, 157
McKinsey 304
Meccano 219
memory 72, 79
Metallica 88
Michelangelo 272
Microsoft 202
Milne 105
mind maps 143, 145
Miranda 16
Monet 32
Morphology strips 152
motivation 58, 59, 60
Mozart 52
MRI 27
multi-voting 165
Murakami 47
Musk 289

N

NASA 182
Natural work teams 203
Nerdery 222
NESTA 278, 280, 304
neuroscience 306
Newton 16, 22, 82
Nietzsche 68
novelty 32
Novelty 29

O

Oasis 63, 65
Observational Approaches 163
Occam's razor 236, 266
Orange 222
Osborn Parnes Creative Problem-
    Solving Model 256

P

Parker 261
Pasteur 39
Pemberton 93
Perception 71
Perot 173
Perry 290

perseverance 103
personality inventories 232
Peters 163
Picasso 42, 47, 57, 59, 77
Picture Associations 147
Pink 92, 218
Pixar 50, 177, 185, 222
Poincaré 38
polymaths 37, 54, 298
preparation 38
Pressfield 91, 101
Prince of Serendip 154
Problem-Solving Process 235
procrastination 91, 101
Proctor and Gamble 187, 189
Pyke 122

Q

Quaternion theory 40

R

Radiohead 42, 163
Random Word Association 114
randomness 157
Rank 57
Rawlinson 105, 106, 127
red teaming 184
Red teaming 274
reframing 139, 140, 146
relaxation 37
research 158
reversal 139
Reversal 138
risky shift 211, 232
Robinson 23, 297
Rock 88
role-playing 134
root cause analysis 236
Root cause analysis 252
Rowe 128
Rowling 93

S

Sadler-Smith 40
Saville 73
Sawyer 29, 31, 61

Schawlow 59
Seek 272
Seelig 51, 172, 287, 290
selection grid 164
Selection of Ideas 164
self-limiting beliefs 98
Semantic Techniques 136
serendipity 157
Shakespeare 282
Shapero 46
Sheeran 22, 302
Shield Exercise 149
Shiseido 222
Siemens 18
Simon 52
situational context 64
Six Thinking Hats 115, 141
skunk works 187
SMART goals 70, 104
Sony 199, 284
Springsteen 283
Stagekings 276
Stanford University 68
Starbucks 163
Sternberg 22, 34
stickiness 292
Stockdale paradox 94
Story Telling 142
Storyboarding 148
Strauss 257
Sundowners 101

T

Tarantino 30, 42, 49
Taylor 190
Team flex 202
Thatcher 93
Thicke 302
TikTok 304

Toyota 34, 173
Tracy 98
Twain 99
Two Level Thinking 125

U

Uber 284
Unilever 163

V

van der Meer 44
Van Gogh 32, 269
verification 40
Visioning 128, 131
Visual Techniques 143
von Oech 148
Von Oech 159
Vunjak-Novakovic 188

W

Wallace 93
Wallas 37, 38
Watterson 39
WD40 92
Wham O Corporation 278
Wilco 93
wilful blindness 283
Williams 302
Wintle 271
World Economic Forum 18, 304
Wright brothers 42, 265

X

Xerox 19

Y

YouTube 72, 291, 292
Yunus 258